Latin *for* Bird Lovers

Latin *for* Bird Lovers

Over 3,000 bird names explored and explained

ROGER LEDERER & CAROL BURR

Timber Press
PORTLAND · LONDON

First Published in the United States and United Kingdom
2014 by Timber Press, Inc.

The Haseltine Building 6a Lonsdale Road
133 S.W. Second Avenue, London NW6 6RD
Suite 450 timberpress.co.uk
Portland, Oregon
97204-3527
timberpress.com

Copyright © 2014 by Quid Publishing. All rights reserved.

Conceived, designed, and produced by Quid Publishing
Printed and bound in China by 1010 Printing Group Limited
Page design by Lindsey Johns

ISBN 13: 978-1-60469-546-5

Catalog records for this book are available from the Library
of Congress and the British Library.

To the individuals and organizations that help us to
understand and appreciate birds and their unique
contribution to a sustainable world

Contents

Preface	8
How to Use This Book	9
A Short History of Binomials	10
Latin for Bird Lovers	12
An Introduction to the A–Z Listings	15

Dohrn's Thrush-Babbler (p. 107)

THE A–Z LISTINGS OF LATIN BIRD NAMES

A	from Aalge to Aythya	16
B	from Bacchus to Buthraupis	29
C	from Cabanisi to Cyrtonyx	39
D	from Dactylatra to Dumetia	56
E	from Eatoni to Exustus	63
F	from Fabalis to Fuscus	71
G	from Gabela to Gyps	83
H	from Haastii to Hypoxantha	96
I	from Ianthinogaster to Ixos	109
J	from Jabiru to Jynx	116
K	from Kaempferi to Kupeornis	118
L	from Labradorius to Lyrurus	122
M	from Macgillivrayi to Myzornis	134
N	from Naevius to Nystalus	148
O	from Oatesi to Oxyura	156
P	from Pachycare to Pyrrhura	164
Q	from Quadragintus to Quoyi	177
R	from Rabori to Rynchops	180
S	from Sabini to Syrmaticus	187
T	from Tabuensis to Tyto	200
U	from Ultima to Ustulatus	207
V	from Validirostris to Vultur	210
W	from Wagleri to Woodfordi	213
X	from Xanthocephalus to Xiphorhynchus	214
Y	from Yarrellii to Yunnanensis	215
Z	from Zambesiae to Zosterops	218

Buteo solitarius,
Hawaiian Hawk (p. 163)

Jynx ruficollis,
Red-throated Wryneck (p. 117)

GENUS PROFILES

Amazona	23
Anas	24
Apteryx	27
Cinnyris	51
Columba	52
Corvus	54
Erithacus	67
Falco	72
Gavia	85
Halcyon	97
Haliaeetus	98
Lanius	123
Melanerpes	139
Meleagris	140
Otus	161
Passer	166
Phoenicopterus	170
Turdus	206
Vanellus	211
Zosterops	219

John Gould
1804–1881 (p. 20)

FAMOUS BIRDERS

John Gould	20
David Lambert Lack	46
Christian Jouanin	76
Phoebe Snetsinger	94
Phillip Clancey	104
James Bond	114
Louis Agassiz Fuertes	132
Konrad Lorenz	154
Alexander F. Skutch	178
Margaret Morse Nice	198
Alexander Wilson	216

BIRD THEMES

Avian Adaptations	34
Bird Beaks	58
The Color of Birds	88
Feathers	120
Bird Songs and Calls	146
Common Names	162
Migration	190
Foraging	208

Glossary	220
Bibliography	222
Credits and Acknowledgments	223

Calypte anna,
Anna's Hummingbird (p. 25)

PREFACE

For most birdwatchers, a good field guide—*The Birds of Western North America*, *The Birds of Europe*, *The Birds of Australia*, the birds of wherever—is sufficient to identify birds in the field. These guides all come in a similar format: Next to the illustration of the species is its common name, typically in larger, bold type (e.g. Desert Lark), while the italicized scientific name (*Ammomanes deserti*) is smaller and lighter. Because birders are generally not interested in the taxonomic or evolutionary relationships of birds, scientific names seem to be of little use.

While most diving ducks have the genus name *Aythya* and most dabbling ducks *Anas*, bird lovers tend to refer to them as dabblers and divers. Although birdwatchers call all large predatory birds raptors, it is just a handy group name for eagles, hawks, and owls. *Empidonax* flycatchers that are difficult to identify are sometimes called "empees," a shortened scientific name that has become a common name.

Scientific names, binomials, are used by scientists to define the exact evolutionary relationships of birds. Using Greco-Latin terms, these names are standardized across the world and are mostly descriptive. If the birdwatcher takes the time to look at these names, they will begin to notice interesting patterns and discover relationships between birds they may not have known about before. For example, there are several genera (plural of genus) of New World sparrows, such as *Spizella*. The scientific name for the American Tree Sparrow, *Spizella arborea*, shows that it is more closely related to *Spizella passerina*, the Chipping Sparrow, than it is to *Chondestes grammacus*, the Lark Sparrow, even though they are all called sparrows.

Most scientific names are at least partly descriptive, such as *Corvus brachyrhynchos*, the American Crow, meaning a short-beaked crow (*brachy*, short, *rhynchos*, beak, *corvus*, crow). *Aix galericulata* comes from *Aix*, duck, *galer*, cap, and *cul-*, little, as in the Mandarin Duck, with a swept-back head crest. Some birds were named to honor an influential person such as an ornithologist, naturalist, politician, or royalty, as in *Estrilda kandti*, Kandt's Waxbill, after Richard Kandt, physician and explorer, and from the German *Wellenastrild*, Waxbill. Other names may describe the place the bird was first found, its color, or behavior. Occasionally, it includes the name of a mythical god, goddess, or creature. You may find that a little bit of research into scientific names opens up a whole new way of looking at and understanding birds.

Latin for Bird Lovers is not only about the origin of scientific names. We also try to explain a little about how and why birds are named and occasionally add tidbits of information about the birds themselves. The book can be picked up and read or referred to in any order in bits and pieces, like a typical dictionary or encyclopedia.

"But true birders, who used to be called bird-watchers, occupy, with other amateur naturalists, a small bywater of the scientific mind in which the naming of things is an overriding hunger. It might better be called bird-naming than bird-watching. The more names, the more finely the distinctions are drawn, the better."

—James Gorman, New York Times, Oct 22, 2002

How to Use This Book

ALPHABETICAL LISTINGS
Scientific terms appear alphabetically throughout for easy reference. For a more detailed explanation, see the Introduction to the A–Z Listings on p. 15.

GENUS PROFILE PAGES
Dotted throughout the book, feature pages examine the interesting characteristics of a particular genus of birds.

The scientific term.

A guide to pronunciation is provided and capital letters indicate where the emphasis should fall.

Aalge AL-jee
Type of auk, from the Danish, as in *Uria aalge*, the Common Murre or Guillemot

An example of a bird name that features the Latin term is given, along with the bird's common name.

LATIN IN ACTION
Feature boxes connect the history of scientific names to individual birds or groups of birds.

FAMOUS BIRDERS
Read the stories of men and women whose enthusiasm for bird-watching has led them around the world, and whose observations have contributed to our understanding of our feathered friends.

BIRD THEMES
In these pages some of the fascinating physical attributes and behaviors of birds are looked at in depth, highlighting any correlation between these aspects and their scientific names.

A Short History of Binomials

A "binomial" is a double name. It is part of a formal system of taxonomy for naming all living things. Carl Linnaeus, a Swedish physician, botanist, and zoologist of the eighteenth century, is considered the father of taxonomy and the system of binomial nomenclature, in which his name is rendered *Carolus linnaeus*.

All living organisms are classified according to their evolutionary relationships and through the classification scheme they are given an individual binomial or scientific name that consists of a genus and species. Human beings, for example, are *Homo sapiens*, meaning they belong to the genus *Homo*, along with other now extinct species such as *Homo habilis* and *Homo erectus*, but are their own specific group, *sapiens*. In usage, the genus, in this case *Homo*, is always capitalized and either underlined or italicized. The species name or specific epithet, *sapiens*, is either underlined or italicized, but never capitalized. And the word "species," as used in biology, is both singular and plural. ("Specie" is incorrect as it means "coin.") Species is often used interchangeably with the scientific name. Classifications change over time as new information develops, but it is a slow and thorough process so the classification scheme is quite stable.

A species is generally defined as a group of organisms that can interbreed and produce viable offspring. *Anas platyrhynchos*, the Mallard, cannot interbreed with *Melanerpes cactorum*, the White-fronted Woodpecker, or even with the more closely related *Anas strepera*, the Gadwall. The species concept continues to evolve with new genetic tools, and there are minor exceptions where different species do hybridize, but the system continues to be useful.

When Linnaeus created the binomial system, "New" Latin was used in Western Europe as the common language of science, and scientific names were in Latin or Greek. Scientific naming is governed by international codes, such as the International Code of Zoological Nomenclature (ICZN) for animals, and the International Code of Nomenclature (ICN) for algae, fungi, and plants.

You might see three names, a trinomial, as in the Red-shouldered Hawk, *Buteo lineatus,* which has five subspecies, *B. l. lineatus*, *B. l. elegans*, *B. l. alleni*, *B. l. extimis*, and *B. l. texanus*. In the hierarchy of taxonomy, only one name below the binomial, a "subspecies," is allowed to denote different color forms or geographic races of a particular species. Hypothetically, all these subspecies can interbreed, but it may not happen if their ranges do not overlap. So whilst subspecies is a somewhat slippery concept, it is useful in delineating populations with distinguishing characteristics.

Carl Linnaeus
(1707–1778)

Linnaeus's classification scheme had an enormous effect on the naming of organisms, now strictly regulated by international rules.

Thus, binomial nomenclature provides a global identification of particular bird species and defines their relationships to other birds. DNA studies over the past couple of decades are refining those relationships and name changes will continue to occur.

Where do the names come from?

- Latin and Greek and occasionally other languages, as in *Anas platyrhynchos*, which comes from the Latin *Anas*, duck, and from the Greek *platy*, flat, and *rhynchos*, bill. *Gavia immer* is the Great Northern Loon or Diver, *Gavia* deriving from Latin, meaning ravenous sea bird, and *immer*, Swedish for ember-goose, referring to the dark color of the bird.
- Names of people, often ornithologists or naturalists, and, in practice, names of people other than the person doing the naming. The White-eared Bronze Cuckoo, *Chrysococcyx meyerii*, was named after Adolf Meyer, a German anthropologist and ornithologist of the late nineteenth and early twentieth century.
- Names of places, as in *Tangara florida*, the Emerald Tanager.
- Local names, like Hoopoe, an onomatopoetic name based on the bird's call. These common names become part of the scientific name, in this case, the Greek *epops*, in *Upupa epops*.
- Descriptions of the bird's color, shape, or behavior such as *Bicinctus*, twice encircled, or banded, as in *Treron bicinctus*, the Orange-breasted Green Pigeon. The Red-headed Myzomela or Honeyeater, *Myzomela erythrocephala*, is from the Greek, *muzao*, suck in; *meli*, honey, Latin, *erythro*, red, and *cephala*, head.
- Odd appellations, such as *Aerodramus fuciphagus*, from *fuci*, seaweed, and *phagus*, eater of, as in the Edible-nest Swiftlet. The species name derives from a Chinese story of the birds swooping down into the ocean to collect material for their nests, which are made almost exclusively of saliva, not seaweed.

Tangara florida, Emerald Tanager (p. 78)

On the whole, the scientific—binomial—names of birds are descriptive in one way or another. More importantly, they can definitively designate each particular species of bird in one language that is officially recognized around the world.

One can only imagine the confusion if birds were identified only by their common names. Instead of *Anas platyrhynchos* all over the northern hemisphere, we'd have Mallard, Canard Colvert, Anade Real, Stokente, Wilde End, Germano Reale, Stokkand, Ma-gamo, Pato-real, and others. Clearly unworkable. So scientific names are valuable, although common names are used most often by birdwatchers.

But since common bird names have caused confusion in the past, the American Ornithologists' Union and British Ornithologists' Union have made some common English names for American and British birds official.

As in all science, continually gathered new information changes taxonomic relationships, and the scientific and common names along with them. So, although we tried to include the most current and accurate scientific names, based on the International Ornithologists' Union's IOC World Bird List, there is no guarantee that they will be accurate tomorrow.

Latin for Bird Lovers

As in every scientific discipline and related endeavors, words derived from the Latin form the core language of the discipline, although derivations from Greek and shared Indo-European roots figure strongly as well. Ornithology and bird-watching are perfect examples. The scientific names of birds define the relationships among some 10,000 species and are typically descriptive. The genus and species name may describe the birds' color, pattern, size, or parts of the body; the name of an ornithologist or other person; where it is found; its behavior; or some characteristic that may not make sense now but did in the eye of the person who named it. In any case, it is often interesting. For example, *Falco mexicanus*, the scientific name of the Prairie Falcon, obviously means a falcon from Mexico. Less obvious, perhaps, is *Anas acuta*, the Northern Pintail, whose scientific name means "sharp duck," referring to the male bird's tail.

Bird enthusiasts don't often pay much attention to scientific names, but bird feather anatomy such as "superciliary" and "auricular" are crucial to identification, as is "furcula" to those banding (or ringing) birds and estimating fat stores on the birds. "Pelagic" is a term not known to most people but often used by ocean-going birders.

We hope this book will open your mind to some scientific and everyday terms that have derived from the Latin and make your bird-watching activities all the more fascinating.

Biological classification is based on the work of Carl Linnaeus, who grouped species on the basis of shared physical characteristics. Darwin's classifications based on evolutionary descent increased the consistency of classification. Now, phylogenetic groupings created by taxonomy and DNA data in addition to morphology are used. Interestingly, these new methods have corroborated much of the anatomical and morphological classifications.

The major taxonomical classifications are class, order, family, genus, and species. Birds belong to the Class Aves and are grouped into 27 orders, all of which end in -iformes, such as Passeriformes (songbirds) and Gaviiformes (loons or divers). Each order contains one or more family, ending in -idae, such as Paridae (titmice). The focus of this book is on the genus and species, the most specific groupings. The genus is always capitalized and italicized; the species is always lower case and italicized; e.g. *Passer domesticus*, the House Sparrow. While taxonomists in all biological fields disagree to some degree about classifications schemes, those in ornithology generally agree.

Anas acuta, Northern Pintail (p. 17)

Gavia immer,
Great Northern Loon or Diver (p. 85)

Because of the loon's eerie call, the name has come to mean crazy and is mistakenly associated with our lunar satellite.

Pronunciation of the names in this book is based on New Latin, the form that developed after the Renaissance (around 1500) for scientific nomenclature, particularly Greco-Latin taxonomic nomenclature of biology.

Unlike classical Latin, New Latin varies from region to region, country to country, and there is no international agreement about how scientific names should be pronounced. Since this book is in English, we are using that language to shape our pronunciations. Our main goal is to give you a workable pronunciation that is true to the etymology of the term—both Greek and Latin. In addition, we are using English pronunciations of places to determine the best management of vowels, consonants, and accents.

Beyond the regional differences in the pronunciation of scientific names, New Latin differs from classical Latin in its handling of vowels, consonants, and accents. For example, there is no th- sound in classical Latin; it is, however, generally used in New Latin but only in its unvoiced form, as in theater. The only exception to this rule is in proper names, such as a person's name (Thomas) or a place name (River Thames). Many birds are named after people, usually not the person doing the naming. These names are "Latinized" to create the binomial.

For example, Audubon named a wren after English ornithologist Thomas Bewick. This bird's common English name is Bewick's Wren, but the Latin name is *Thryomanes bewickii* (Greek, *thruon*, reed, and *manes*, very fond of) to describe the preferred habitat of the bird. Bewick's name becomes *Bewickii, -ii* to show possession, giving us a pronunciation of *be-WIK-ee-eye*. While the accent is generally put where it exists in the language of the name, the Latin possessive form sometimes shifts that accent and accounts for some of the variations in pronunciation between regions and countries.

Passer ammodendri,
Saxaul Sparrow (p. 166)

The songbirds, order Passeriformes, comprise the largest taxonomic group of birds, containing about 52 percent of all bird species.

Hylonympha macrocerca,
Scissor-tailed Hummingbird (p. 108)

An Introduction to the A–Z Listings

This book is intended to be an interesting and enjoyable jaunt into the world of scientific bird names. It is not definitive. That would take a much bigger book and a much greater interest in the derivation of names. The *Helm Dictionary of Scientific Bird Names* by James Jobling is the closest to a definitive source of bird names (about 20,000 of them on 432 pages) that we have found; we used the International Ornithologists' Union's IOC World Bird List as the final arbiter on the accuracy of common and scientific bird names.

Here follows, in alphabetical order, over 3,000 scientific names of birds, either a genus or species. The term appears first, then a guide to the pronunciation from New or Scientific Latin, followed by the definition of the term, or its parts. An example of a scientific name of which the word is a part is also given. For example:

> Caeruleirostris *see-roo-lee-eye-ROSS-tris*
> *Caerul*, blue, and *rostris*, bill or beak, as in *Loxops caeruleirostris*, the Akeaee, a honeycreeper with a blue bill

In this example, the words are Latin derived and the two parts of *caeruleirostris* are defined. As added clarification, as the common name is not helpful, it is noted as a honeycreeper with a blue bill. Most of the scientific names have clear meanings, although why they were chosen for a particular bird may sometimes not be clear. In the interest of space, explanations of why a bird was given a particular scientific name are provided only where we think the reader's curiosity might be piqued.

For simplicity, not all the alternative endings, genders, or cases of each term are listed, as the definitions of words like *leptoptila* and *leptotilos* are essentially identical (*lepto*, thin, slender, and *ptilo*, wing).

The language of derivation of the scientific name is noted if it is other than Latin. If no language is noted, it is from the Latin. It is worth noting again that Latin (including Classical, Modern, Scientific, Late, Medieval, Renaissance) and Greek evolved from shared Indo-European roots, producing significant overlap in etymology. Therefore, as often as possible, we have used the Latin spellings and pronunciations because they comprise the main basis for New (scientific) Latin. We have also chosen the most applicable explanation to elucidate the naming of birds.

Loxops caeruleirostris
Akekee, a honeycreeper (p. 39)

A

Aalge AL-jee
Type of auk, from the Danish, as in *Uria aalge*, the Common Murre or Guillemot

Abbotti AB-bot-tye
After William Louis Abbott, American doctor and naturalist, as in *Papasula abbotti*, Abbott's Booby

Abeillei a-BEL-eye
After M. Abeille, French collector and naturalist, as in *Icterus abeillei*, the Black-backed Oriole

Aberrans AB-ber-ranz
Unusual, different, as in *Cisticola aberrans*, the Lazy Cisticola; denotes use of an unusual habitat for this group of birds

Aberti AL-bert-eye
After American ornithologist James William Abert, as in *Melozone aberti*, Abert's Towhee

Abnormis ab-NOR-mis
Ab, away, and *normis*, usual; so, abnormal, as in *Sasia abnormis*, a very small woodpecker, the Rufous Piculet

Abroscopus a-bro-SKO-pus
Greek, *abro*, delicate, dainty, and *skopus*, sentry, guard or goal, as in *Abroscopus superciliaris*, the Yellow-bellied Warbler

Aburria a-BUR-ree-a
From *abhorrere*, to abhor, not inclined to, bored, as in *Aburria aburri*, the Wattled Guan; applied to the bird for its habit of resting for the better part of a day

Abyssinicus, -a a-bis-SINK-us/a
After East Africa, specifically Abyssinia, now Ethiopia, as in *Asio abyssinicus*, the Abyssinian, or African Long-eared, Owl

Acadicus a-KAD-ih-kus
After Acadia, a region in Canada, as in *Aegolius* (owl) *acadicus*, the Northern Saw-whet Owl

Acanthagenys a-kan-tha-JEN-is
Greek, *akanthos*, from *ake*, point, thorn, and *genys*, jaw, as in *Acanthagenys rufogularis*, the Spiny-cheeked Honeyeater

Acanthis a-KAN-this
Greek, *akanthos*, from *ake*, point, thorn; Zeus and Apollo changed Acanthus into a finch after Acanthus was killed by his father's horse, as in *Acanthis flammea* (red flag), the Common Redpoll

Acanthiza a-kan-THY-za
Greek, *akanthos*, from *ake*, point, thorn, and *zo*, to live, as in *Acanthiza inornata*, the Western Thornbill

Acanthorhynchus a-kan-tho-RINK-us
Greek, *akanthos*, from *ake*, point, thorn, and Latin, *rhynchus*, bill, as in *Acanthorhynchus superciliosus*, the Western Spinebill

Acanthornis a-kan-THOR-nis
Greek, *akanthos*, from *ake*, point, thorn, and *ornis*, bird, as in *Acanthornis magna*, the Scrubtit

Accipiter ak-SIP-ih-ter
To take, grasp, receive. A group of predatory birds, forest dwelling and diurnal, such as the *Accipiter gentilis*, Northern Goshawk

Aceros a-SER-os
Greek, *a*, without, and *ceros*, horn, as in *Aceros corrugatus*, the Wrinkled Hornbill

Acridotheres a-kri-do-THER-eez
Greek, *akridis*, locust, and *therao*, hunt, as in *Acridotheres tristis*, the Common Myna

Sasia abnormis, Rufous Piculet

Acrobatornis *a-kro-ba-TOR-nis*
Greek, *acrobat*, gymnastic performer, and *ornis*, bird, as in *Acrobatornis fonsecai* (hunter), the Pink-legged Graveteiro, an acrobatic bird that can hang upside down while foraging

Acrocephalus *a-kro-se-FAL-us*
Acro, acrobatic and *cephala*, head, as in *Acrocephalus agricola*, the Paddyfield Warbler

Actenoides *ak-ten-OY-deez*
Greek, *aktis*, ray or beam, and *oides*, resembling, as in *Actenoides concretus*, the Rufous-collared Kingfisher

Actinodura *ak-tin-o-DOO-ra*
Greek, *aktis*, ray or beam, and *oura*, tail, as in *Actinodura egertoni*, the Rusty-fronted Barwing, with a pointed tail

Actitis *ak-TY-tis*
Greek, coast dweller, as in *Actitis macularius*, the Spotted Sandpiper

Actophilornis *ak-to-fil-OR-nis*
Greek, *aktis*, ray or beam, *philos*, like or love, and *ornis*, bird, as in *Actophilornis africanus*, the African Jacana; means sun-loving bird

Acuminata *a-koo-min-AH-ta*
Sharp or tapering, as in *Calidris acuminata*, the Sharp-tailed Sandpiper

Acuta *A-KOO-ta*
Sharp, as in *Anas acuta*, to describe the pointed tail of the Northern Pintail

Acutipennis *a-koo-tih-PEN-nis*
Acuta, sharp, and *penna*, feather, as in *Chordeiles acutipennis*, the Lesser Nighthawk

Adelberti *a-DEL-bert-eye*
After Adelbert Fényes de Csakaly, a noted physician, entomologist, and ornithologist, as in *Chalcomitra adelberti*, the Buff-throated Sunbird

Adorabilis *a-do-RA-bil-is*
Adoro, revere, honor, worship, adore, as in *Lophornis adorabilis*, the White-crested Coquette

Adscitus *ad-SHE-tus*
Adopt, approve, as in *Platycercus adscitus*, the Pale-headed Rosella. Named and renamed in the late 1700s, the etymology of the species name is unknown

Actenoides concretus,
Rufous-collared Kingfisher

Adsimilis *ad-SIM-il-is*
Similar, close, as in *Dicrurus adsimilis*, the Fork-tailed Drongo; the namer thought the drongo looked like a common bird of the UK

Aechmophorus *ek-mo-FOR-us*
Greek, *aikhme*, spear, and *phero*, bearing, as in *Aechmophorus occidentalis*, the Western Grebe; named for its spear-like bill

Aedon *EE-don*
In Greek mythology, after Aedon, wife of Zethus, who was turned into a bird by Zeus, as in *Troglodytes aedon*, the House Wren

Aegithalos *ee-ji-THAL-os*
Greek for tit, a small bird, as in *Aegithalos fuliginosus*, the Sooty Bushtit; tit from Norwegian *titr*, small bird

Aegotheles *ee-go-THEL-eez*
Greek, *aego*, goat, and *theles*, suck, suckle, as in *Aegotheles insignis*, the Feline Owlet-nightjar; members of this family are called "goatsuckers"

Aegypius *ee-JIP-pee-us*
Greek, *aigupios*, vulture, as in *Aegypius monachus*, the Cinereous or Black Vulture

Aeneus *ee-NEE-us*
Brassy or gold in color, as in *Dicrurus aeneus*, the Bronzed Drongo

Aenigma *ee-NIG-ma*
Riddle, as in *Sapayoa aenigma*, the Sapayoa

Aepypodius *ee-pi-PO-dee-us*
Greek, *aipus*, tall, high, and *pous*, foot, as in *Aepypodius bruijnii*, the Waigeo Brushturkey

Aequatorialis *ee-kwa-tor-ee-AL-is*
Equatorial, as in *Momotus aequatorialis*, the Andean Motmot

Aerodramus *eh-ro-DRA-mus*
Greek, *aer*, air, and *dram*, to run, as in *Aerodramus elaphrus*, the Seychelles Swiftlet

Aestiva, -alis *es-TEE-va/es-tee-VAL-is*
Summer, as in *Peucaea aestivalis*, Bachman's Sparrow and *Amazona aestiva*, the Turquoise-fronted Amazon

Aethereus *ee-THER-ee-us*
Greek, *aitherios*, ethereal, as in *Phaethon aethereus*, the Red-billed Tropicbird

Afer *AH-fer*
Used by ancient Romans to refer to what is now Tunisia, as in *Euplectes afer*, the Yellow-crowned Bishop

Affinis *af-FIN-is*
Affinity, similarity, as in *Aythya affinis*, the Lesser Scaup, which is closely related and virtually identical to the Greater Scaup, *Aythya marila*. *Affinis* is the specific epithet of dozens of bird species

Agapornis *a-ga-POR-nis*
Greek, *agape*, love or affection, and *ornis*, bird, as in *Agapornis fischeri*, Fischer's Lovebird

Agelaioides *a-jel-eye-OY-deez*
Greek, *agelaius*, gregarious, and *oides*, resembling, as in *Agelaioides badius*, the Baywing

Agelaius *a-je-LE-us*
Greek, gregarious, as in *Agelaius tricolor*, the Tricolored Blackbird, a colonial nesting species that also winters in large flocks

Agilis *a-JIL-is*
Agile, nimble, quick, as in *Oporornis agilis*, the Connecticut Warbler

Aglaiae *a-GLAY-ee*
Agali, brilliant, splendid, as in *Pachyramphus aglaiae*, the Rose-throated Becard

Agricola *a-GRI-ko-la*
Ager, field, and *cola*, inhabitant, as in *Acrocephalus agricola*, the Paddyfield Warbler

Agriornis *ah-gree-OR-nis*
Agri, agriculture, and Greek, *ornis*, bird, as in *Agriornis micropterus*, the Gray-bellied Shrike-Tyrant

Ailuroedus *eye-loo-ROY-dus*
Greek, *ailur*, cat, and *oidos*, singing, as in *Ailuroedus melanotis*, the Spotted Catbird

Aimophila *eye-mo-FIL-a*
Greek, *aimos*, thicket, and *philos*, to like, as in *Aimophila ruficeps*, the Rufous-crowned Sparrow

Aix *EYKS*
Greek, waterfowl, as in *Aix sponsa*, the Wood Duck

Ajaja *a-JA-ja*
Finnish, to drive, ride, or chase, as in *Platalea ajaja*, the Roseate Spoonbill (see box)

Aix sponsa,
Wood Duck

Alauda *a-LAW-da*
Celtic, great song, as in *Alauda arvensis*, the Eurasian Skylark, noted for its sustained singing while on the wing

Alba, -i, -o *AL-ba/beye/bo*
White, as in *Motacilla alba*, the White or Pied Wagtail

Albatrus *al-BAT-rus*
Probably derived from Portuguese *alcatraz*, pelican, *gha*, a kind of sea eagle, as in *Phoebastria albatrus*, the Short-tailed Albatross

Albellus *al-BEL-lus*
Diminutive of *alba*, as in *Mergellus albellus*, the Smew, a small duck releated to the mergansers

Alberti *AL-bert-eye*
After Prince Albert, husband of Queen Victoria, as in *Crax alberti*, the Blue-billed Curassow

Albescens *AL-bes-sens*
Albus, white and *-escens*, becoming, as in *Calendulauda albescens*, the Karoo Lark

Albicapillus, -a *al-bi-ka-PIL-lus/a*
Albus, white, and *capillus*, hair, as in *Lamprotornis albicapillus*, the White-crowned Starling

Albicaudatus, -a *al-bi-kaw-DA-tus/ta*
Albus, white, and *cauda*, tail of an animal, as in *Geranoaetus albicaudatus*, the White-tailed Hawk

Albiceps *AL-bi-seps*
Albus, white, and *ceps*, head, as in *Vanellus albiceps*, the White-crowned Lapwing

Albicilla *al-bi-SIL-la*
Albus, white, and *cilla*, tail, as in *Haliaeetus albicilla*, the White-tailed Eagle

Albicollis *al-bi-KOL-lis*
Albus, white, and *collis*, throat or collar, as in *Corvus albicollis*, the White-necked Raven

Albidinucha *al-bi-di-NOO-ka*
Albus, white, *idus*, having the nature of, and *nucha*, nape, as in *Lorius albidinucha*, the White-naped Lory

Albifacies *al-bi-FACE-eez*
Albus, white, and *facies*, face, as in *Geotrygon albifacies*, the White-faced Quail-Dove

---

LATIN IN ACTION

The Roseate Spoonbill, *Platalea ajaja*, derives its scientific name from its feeding habits. It is found in shallow coastal waters of the southeastern US, where it walks with its spoon-shaped bill in the water and mud, driving prey in front of it, snatching frogs, crabs, fish, worms, crayfish, and other such creatures. In the process of swallowing their prey, the birds also ingest microorganisms with reddish pigments that give the birds their pink color. This behavior and resulting coloring resembles that of the flamingos. Young nestling spoonbills, fed by their parents via regurgitation, have ordinary-shaped bills that gradually become spoon-shaped over a period of several months.

Platalea ajaja, Roseate Spoonbill

Albifrons *AL-bi-fronz*
Albus, white, and *frons*, forehead, as in *Amazona albifrons*, the White-fronted Amazon

Albigula *al-bi-GOO-la*
Albus, white, and *gula*, gullet as in *Buteo albigula*, the White-throated Hawk

Albilatera *al-bi-la-TER-ra*
Albus, white, and *latera*, side, as in *Diglossa albilatera*, the White-sided Flowerpiercer

Albipectus *al-bi-PEK-tus*
Albus, white, and *pectus*, chest, as in *Pyrrhura albipectus*, the White-breasted Parakeet

John Gould

(1804–1881)

John Gould was born in Dorset, England. His father, a gardener at Windsor Castle, taught him the skills of his trade and Gould eventually secured a position as a gardener at Ripley Castle in Yorkshire. He received little formal education but eventually came to be considered the Father of Australian Ornithology. He was as well known in Europe as Audubon was in America.

Gould also developed skills as a taxidermist and established a taxidermy business in London which stood him in good stead as he built his career as an ornithologist. His contacts with scientists and natural historians led to his obtaining a position as the first curator and preserver of the Zoological Society museum in London.

As curator, Gould had access to all the specimens given to, or collected for, the society. In 1830 he received a collection of birds from the Himalayas, many of which were new to Europe, and turned it into a book, *A Century of Birds from the Himalaya Mountains*, with his new wife Elizabeth doing many of the illustrations. Over the next few years Gould wrote four bird books, one a five-volume Birds of Europe, with beautiful lithographs by Edward Lear. From the age of 20 to 26, Lear created 80 bird portraits for Gould. Many people consider them to be the world's finest ornithological illustrations of the time. Unfortunately, they were mixed in Gould's books with less accomplished work by other artists.

In 1837 Gould met Charles Darwin. Darwin had just returned from the Galapagos and his bird specimens were given to Gould to identify. He realized that the birds Darwin thought were separate species were actually varieties of the same species adapted to the conditions on individual islands. Gould's analysis led to an important step in Darwin's thinking on what became his theory of evolution by natural selection. Gould's work on the birds was included in *Zoology of the Voyage of the H.M.S. Beagle*, and published between 1838 and 1842 with Darwin as editor.

In 1838 Gould and his wife sailed to Australia, wanting to be the first to compile a book on the avifauna of the country. Returning to England in

Dendrocitta vagabunda,
Rufous Treepie

The Rufous Treepie, like other members of the crow family Corvidae, is omnivorous and has adapted to a variety of habitats.

1840, Gould produced *The Birds of Australia*, consisting of 600 illustrations in seven volumes and describing over 300 new bird species. After his wife died in childbirth in 1841, Gould went on to publish *A Monograph of the Trochilidae or Humming Birds* (1849–61), *The Mammals of Australia* (1845–63), *Handbook to the Birds of Australia* (1865), *The Birds of Asia* (1850–83), *The Birds of Great Britain* (1862–73), and *The Birds of New Guinea and the Adjacent Papuan Islands* (1875–88). He was certainly the most prolific ornithological author of his time, producing 41 publications on birds, containing almost 3,000 illustrations created by his wife and others. He was also a superb painter, creating prints that were in great demand.

Some researchers think that Gould himself did the original sketches for all the plates and that Elizabeth Gould, Edward Lear, and others did the hand coloring and lithography. While Gould was not the illustrator of his books, he was skilled in producing quick sketches of dead birds in the field from which the artists created finished pieces. The results were beautiful books and works of art in every detail. For example, he captured the iridescence of hummingbirds by applying gold leaf under the watercolors.

Campylopterus largipennis, Gray-breasted Sabrewing

For much of his professional career Gould was fascinated by hummingbirds and managed to accumulate a collection of 320 species. In 1851 he displayed this collection at the Great Exhibition in London, a precursor to World's Fairs. In spite of his fascination he had never observed a live hummingbird until he traveled to the US in 1857 and saw a Ruby-throated Hummingbird in Bartram's Gardens in Philadelphia. He captured a few and tried to bring them back alive to England, but needing special care, they lasted only a few weeks.

In 1909 the Gould League of Bird Lovers was founded in Victoria, Australia, to promote environmental education; today it continues to be very active throughout the country. In 1976 the Australian Post honored him with a postage stamp bearing his portrait. In 2009, a series of birds from his Birds of Australia was featured in another set of stamps. Also, at least two dozen birds have been named after Gould: Gould's Petrel, Gould's Bronze Cuckoo, Gould's Frogmouth, Gould's Parrotbill, Gould's Sunbird, and the Gouldian Finch, for example.

"Man's constant companions in every outdoor occupation, cheering him with their presence and their songs, and often affording him a principal means of subsistence, it is little wonder that the habits and instincts of birds should be a favourite one with most persons…"

John Gould, The Birds of Great Britain *(1873)*

Albipennis *al-bi-PEN-nis*
Albus, white, and *pennis*, tail or quill, as in *Petrophassa albipennis*, the White-quilled Rock Pigeon

Albogularis *al-bo-goo-LAR-is*
Albus, white, and *gula*, throat, as in *Phalcoboenus albogularis*, the White-throated Caracara

Albolarvatus *al-bo-lar-VA-tus*
Albus, white, and *larvare*, bewitch/enchant, as in *Picoides albolarvatus*, the White-headed Woodpecker; of 22 species of woodpeckers in North America, this is the only one with a white head, making it particularly unusual and captivating

Albonotatus *al-bo-no-TA-tus*
Albus, white, and *notatus*, marked, as in *Buteo albonotatus*, the Zone-tailed Hawk

Albus *AL-bus*
White, as in *Chionis albus*, the Snowy Sheathbill

Alca *AL-ka*
From Icelandic or Norwegian, *auk*, as in *Alca torda*, the Razorbill

Alcedo *al-SEE-doe*
Kingfisher, as in *Alcedo atthis*, the Common Kingfisher. "Kingfisher" refers to the excellent fishing skills of riparian species

Aleadryas *al-ee-a-DRY-as*
Alea, game, and *dryas*, wood-nymph, as in *Aleadryas rufinucha*, the Rufous-naped Whistler

Alectoris *a-lek-TOR-is*
Greek *alektoris*, cock, as in *Alectoris graeca*, Rock Partridge

Aleuticus *a-LOY-ti-kus*
After the Aleutian Islands, as in *Onychoprion aleuticus*, the Aleutian Tern

Eremophila alpestris,
Horned or Shore Lark

Alexandrae *a-lex-AN-dree*
After Alexandra of Denmark, Queen of the United Kingdom and Empress of India and the wife of King Edward VII, as in *Polytelis alexandrae*, Princess Parrot

Alexandrinus *a-lek-zan-DRY-nu*
After Alexandria (Egypt), as in *Charadrius alexandrinus*, the Kentish Plover

Alle *AL-le*
Norse, small, as in *Alle alle*, the Little Auk

Allenia *AL-len-ee-a*
After Joel Allen, American ornithologist, as in *Allenia fusca*, the Scaly-breasted Thrasher

Alopex *AL-o-pecks*
Greek, fox, a cunning person, as in *Falco alopex*, the Fox Kestrel

Alopochen *al-o-PO-ken*
Greek, *alopex*, fox, and *chen*, goose, as in *Alopochen aegyptiaca*, the Egyptian Goose

Alpestris *al-PES-tris*
Of high mountains; the Horned or Shore Lark's name *Eremophila alpestris* means "love of lonely places in the mountains"

Alphonsionis *al-fon-see-OWN-is*
After Alphonse Milne-Edwards, a French physician and ornithologist, as in *Sinosuthora alphonsiana*, the Ashy-throated Parrotbill

Alpina *al-PINE-a*
Alpine, as in *Calidris alpina*, the Dunlin

Altiloquus *al-ti-LOW-kwus*
Altus, high, and *loquus*, voice, as in *Vireo altiloquus*, the Black-whiskered Vireo

Altirostris *al-ti-ROSS-tris*
Altus, tall, deep, and *rostris*, bill or beak, as in *Turdoides altirostris*, the Iraq Babbler

Amazilia, -zonia *a-ma-ZIL-ee-a/a-ma-ZON-ee-a*
After the Amazon region, as in *Amazilia lactea*, the Sapphire-spangled Emerald (hummingbird)

Amblyornis *am-blee-OR-nis*
Greek, *amblus*, blunt, and *ornis*, bird, as in *Amblyornis flavifrons*, the Golden-fronted Bowerbird. Compared with other genera of Bowerbirds, *Amblyornis* species are rather dull colored

Amazona

About 30 parrot species native to the New World and distributed from the Caribbean to South America are of the *Amazona* (*am-a-ZONE-a*) genus. They are well known around the world for their ability to mimic human voices, to manipulate objects with their feet, and to adjust to captivity, making them in demand as domestic pets. There are over 11 million pet birds in the US, 75 percent of them some kind of parrot. Because the *Amazona* species are among the most popular, there has been a significant decline in many of these species' populations, both through the legal and illegal pet trade. Over 60 percent of parrots caught in the wild for the pet trade die before they reach market.

Their desirability is understandable as their personalities and colors are very appealing, but the habits of wild parrots are not well known. They are difficult to catch as they inhabit the canopy of tall trees and frustrate researchers by sitting there for long periods of time. If they are caught and banded/ringed, they pry off the band/ring with their strong beaks.

Parrots typically eat nuts, fruit, nectar, and occasionally insects or other arthropods. Their zygodactyl feet (the second and third toes are forward while the first and fourth face rearward) are adapted for grasping food objects and their jaws are made for opening the hardest nuts and fruits. The upper jaw is hinged at the skull and curved downward, allowing it to exert considerable pressure against the sharp edges of the flat lower jaw. Touch receptors in the bill allow the bird to manipulate food items to the proper position for being cracked open. To open a Brazil nut the bill has to exert 1,400 pounds per square inch (9,653 kilopascals); this is far more force than it would take to break your finger!

Amazona festiva,
Festive Amazon or Parrot

There are a number of interesting names in the *Amazona* genus. *A. farinosa* (from the Latin *farina*, flour) is called the Mealy Amazon or Parrot because its back and nape appear to have been covered with a layer of flour (meal). The Festive Amazon or Parrot, *A. festiva*, is one of the most festively colored of the genus, although it has a lot of competition.

Amazona violacea,
Guadeloupe Parrot

The Guadaloupe Parrot became extinct in the late eighteenth century but a description of it indicates that the head, neck, and upper thorax were violet colored.

Anas

The Latin for duck is *Anas* (*AN-as*). This genus of waterfowl encompasses about 45 species in the subfamily Anatinae, known as the dabbling ducks for their habit of tipping their heads down and their tails up while feeding on the water. These are ducks like mallards, wigeons, teals, pintails, and shovelers. The most well-known of this genus is certainly the Mallard, *A. platyrhynchos* (Greek, *platys*, broad, and *rhynchos*, bill), with its wide, flat bill; Mallards are found naturally almost everywhere in the temperate and subtropical northern hemisphere and have been introduced elsewhere. With their long, rounded, relatively flat bills with a hook-like nail at the end, dabbling ducks are well adapted for dipping from the surface and searching for food on the bottoms of shallow waters. Inside the edges of the bill are lamellae, comb-like structures that serve as sieves. A mouthful of mud and water is taken in the mouth and filtered for food items like insects and seeds. The especially wide bill that gives the Northern Shoveler its name, *A. clypeata* (Latin *clypeatus*, shield-bearing), may have over 200 lamellae.

Ducks, from the Old English ducan, to duck or dive, comprise perhaps the most recognizable group of birds. Ducks are waterfowl, like geese and swans, but unlike the others, they are sexually dimorphic, the males being much more colorful than the females, and for good reason. On the wintering grounds, the male, in his showy courtship plumage, attracts a drab female; they pair up and migrate to their breeding grounds, where the female excavates a depression in the ground and lines it with nearby grasses. She will lay anywhere from one to a dozen eggs or more and start incubation only after they are all laid, so they all hatch at once and follow the mother around to learn duck survival skills. Her drab plumage serves her well as camouflage during this process.

An interesting urban myth about ducks is that their quack does not produce an echo. Ridiculous on the face of it, this belief has been disproven in the lab.

Anas platyrhynchos, Mallard

Americana a-mer-i-KAN-a
Of America, as in *Recurvirostra americana*, the American Avocet

Ammodramus am-mo-DRA-mus
Greek, *ammos*, sand, and *dramos*, to run, as in *Ammodramus savannarum*, the Grasshopper Sparrow

Amoena, -us a-MOY-na/nus
Lovely, beautiful, as in *Passerina amoena*, Lazuli Bunting

Ampeliceps am-PEL-ih-seps
Ampelos, coiling like a vine, and *ceps*, headed, as in *Ampeliceps coronatus*, the Golden-crested Myna

Amphispiza am-fi-SPY-za
Greek, *amphi*, having two alternates, and *spiza*, finch, as in *Amphispiza bilineata*, the Black-throated Sparrow, originally considered a finch

Anas AN-as
Greek, duck, as in *Anas platyrhynchos*, the Mallard

Anhinga an-HIN-ga
From Tupi Indian (Brazil) language, as in *Anhinga anhinga*, the Anhinga

Anisognathus an-ih-sog-NA-thus
Greek, *aniso*, unequal, and *gnathos*, jaw, as in *Anisognathus notabilis*, the Black-chinned Mountain Tanager

Anna AN-na
After Princess Anna d'Essling, Duchess of Rivoli, as in *Calypte anna*, Anna's Hummingbird

Anomalospiza an-om-o-lo-SPY-za
Greek, *anomalos*, odd, and *spiza*, finch, as in *Anomalospiza imberbis*, the Cuckoo-finch

Calypte anna,
Anna's Hummingbird

Anous AH-noos
Greek, silly, stupid, as in *Anous stolidus*, the Brown Noddy, which shows little fear of humans

Anser AN-ser
Goose, as in *Anser anser*, the Greylag Goose

Anthobaphes an-tho-BAF-eez
Greek, *anthos*, flower, *baph*, dye, dip, as in *Anthobaphes violacea*, the Orange-breasted Sunbird

Anthocephala an-tho-se-FAL-a
Greek, *anthos*, flower, Latin, *cephala*, head, as in *Anthocephala floriceps*, the Blossomcrown, a type of hummingbird

Anthonyi an-THONE-ee-eye
After Alfred W. Anthony, American bird collector and ornithologist, as in *Nyctidromus anthonyi*, Anthony's Nightjar

Anthornis an-THOR-nis
Greek, *anthos*, flower, *ornis*, bird, as in *Anthornis melanura*, the New Zealand Bellbird

Anthoscopus an-tho-SKO-pus
Greek, *anthos*, flower, *skopos*, seeker, as in *Anthoscopus caroli*, the Gray Penduline Tit

Anthracinus An-thra-SYE-nus
Coal black, as in *Buteogallus anthracinus*, the Common Black Hawk

Recurvirostra americana,
American Avocet

Anthus *AN-thus*
Greek, flower, as in *Anthus rubescens* (fresh, reddish), the Buff-bellied Pipit, pipit meaning to chirp. Name perhaps based upon the flower-colored Western Yellow Wagtail, *Motacilla flava*, a member of the same family as the pipit

Antiquus *an-TI-kwuss*
Old, as in *Synthliboramphus antiquus*, the Ancient Murrelet

Aphelocoma *a-fe-lo-KO-ma*
Greek, *apheles*, smooth, and *kome*, hair, as in *Aphelocoma coerulescens*, the Florida Scrub Jay

Apicalis *a-pi-KA-lis*
Tipped, referring to the tail, as in *Moho apicalis*, the extinct Oahu Oo

Aquaticus *a-KWAT-ih-kus*
Aquatic, as in *Rallus aquaticus*, the Water Rail

Aquila *a-KWIL-a*
Eagle, as in *Aquila nipalensis*, the Steppe Eagle

Arachnothera *a-rak-no-THER-a*
Greek, *arachno*, spider, and *thera*, hunt, as in *Arachnothera chrysogenys*, the Yellow-eared Spiderhunter

Arborea *ar-BOR-ee-a*
Tree, as in *Spizella arborea*, the American Tree Sparrow

Archboldia *arch-BOLD-ee-a*
After Richard Archbold, zoologist at the American Museum of Natural History, as in *Archboldia papuensis*, Archbold's Bowerbird

Archeopteryx *ar-kee-OP-ter-iks*
Greek, *Archeo*, ancient, and *pteryx*, wing, as in *Archeopteryx lithographica*, "ancient wing," often considered the "first bird"

Archilochus *ar-kee-LO-kus*
Greek, *archi*, chief, and *lochus*, ambush, as in *Archilochus colubris*, the Ruby-throated Hummingbird, probably because of the bird's territorial behavior

Arctica *ARK-ti-ka*
Northern, as in *Gavia arctica*, the Black-throated Loon or Diver

Arenaria *a-ren-AR-ee-a*
Sand-pit, as in *Arenaria interpres*, the Ruddy Turnstone

Larus argentatus, European Herring Gull

Argentatus *ar-jen-TA-tus*
Adorned with silver, as in *Larus argentatus*, the European Herring Gull

Argus *AR-gus*
Greek, *argos*, the bright one, as in *Argusianus argus*, the Great Argus (pheasant)

Arquata *ar-KWA-ta*
Curved, bow-shaped, as in *Numenius arquata*, the Eurasian Curlew

Asio *AH-see-o*
Little horned owl, as in *Asio otus*, the Long-eared Owl

Assimilis *as-SIM-il-is*
Like, similar, as in *Circus assimilis*, the Spotted Harrier, similar to the Swamp or Marsh Harrier

Ater *AH-ter*
Black, as in *Daptrius ater*, the Black Caracara and *Molothrus ater*, the Brown-headed Cowbird

Athene *ah-THEE-nee*
After Athena, Greek goddess of wisdom, as in *Athene noctua*, Little Owl

Atilla *ah-TIL-la*
From Turkic, original name of Volga River, as in *Atilla rufus*, the Gray-hooded Atilla. Atilla flycatchers are so named because of their aggressive nature, as in Atilla the Hun

Atra *AT-ra*
Black, as in *Myiagra atra*, the Biak Black Flycatcher

Atratus *ah-TRA-tus*
Atra, black, as in *Coragyps atratus*, the American Black Vulture

APTERYX

The genus name *Apteryx* (AP-ter-iks) derives from the Greek *a*, without, lacking, and *pteryx*, wing, although the five species in this genus do in fact have wings, albeit very small and almost unnoticeable. These are the kiwis, the common name derived from the Maori name for the call. The species are the Great Spotted, *A. haastii*, and Little Spotted, *A. owenii*, Kiwis, the Okarito, *A. rowi*, Southern Brown, *A. australis*, and North Island Brown, *A. mantelli*, Kiwis. All are restricted to New Zealand, and New Zealanders are often referred to as Kiwis.

Kiwis are ratites, a group of large flightless birds including the Ostrich, Rhea, Emu, and two cassowaries. There are about 40 species of flightless birds in the world, but ratites are a special group because they do not possess a keel (carina) on their sternum. In flying birds the keel anchors the large flight muscles, but ratites have no keel and only poorly developed breast muscles. Ratite, from the Latin *ratis*, meaning ship, refers to the keel-less sternum that resembles a simple boat. Ratites have no tail, their feathers are primitive with no barbules to hook the barbs together, and there is no preen gland to oil their feathers.

Kiwis are unusual in a number of other ways. They are nocturnal, they have long rictal bristles that serve a tactile function, they probe in the ground with their long bills for worms, and they lay the largest eggs in the world in proportion to their size. About the size of a chicken and weighing 4 to 6 pounds (1.5 to 3.3 kilograms), a kiwi lays an egg that weighs about one quarter of its body weight and is roughly six times the size of the average chicken egg. As you might expect, the female has to eat considerably more than usual during the 30 days the egg is developing.

Apteryx haastii,
Great Spotted Kiwi

Many flightless birds, like the kiwis in New Zealand, evolved on islands with few or no land predators. But with the introduction of cats, weasels, opossums, and the reduction of suitable habitat, populations have declined drastically. Only about 5 percent of kiwi chicks survive in the wild and only in areas where there is some predator control.

Apteryx australis,
Southern Brown Kiwi

Atricapilla *ah-tri-ka-PIL-la*
Atra, black, and *capill*, hair, as in *Sylvia atricapilla*, the Eurasian Blackcap

Atricilla *a-tri-SIL-la*
Atra, black, and *cilla*, tail, as in *Leucophaeus atricilla*, the Laughing Gull

Atricristatus *a-tri-kris-TA-tus*
Atra, black, and *cristatus*, crest, as in *Baeolophus atricristatus*, the Black-crested Titmouse

Atrogularis *aa-tro-goo-LAR-is*
Atra, black, and *gula*, throat, as in *Spizella atrogularis*, the Black-chinned Sparrow

Audouinii *aw-DWIN-nee-eye*
After Jean Victoire Audouin, French naturalist, as in *Ichthyaetus audouinii*, Audouin's Gull

Augur *AW-ger*
Tell the future, as in *Buteo augur*, the Augur Buzzard; perhaps after the augur, the Roman priest who interpreted the gods' wills by studying birds' flight

Aura *AW-ra*
Breeze, air, as in *Cathartes aura*, the Turkey Vulture

Auratus *aw-RA-tus*
Aurata, golden or gilded, as in *Icterus auratus*, the Orange Oriole

Auriceps *AW-ri-seps*
Aurum, gold, and *ceps*, head, as in *Pharomachrus auriceps*, the Golden-headed Quetzal

Auricularis *aw-ri-koo-LA-ris*
Pertaining to the ear, as in *Myiornis auricularis*, the Eared Pygmy Tyrant

Aurifrons *AW-ri-fronz*
Aurum, gold, and *frons*, front, forehead, as in *Ammodramus aurifrons*, the Yellow-browed Sparrow

Auritus *aw-RYE-tus*
Auris, ear, with ears, as in *Phalacrocorax auritus*, the Double-crested Cormorant (see box)

Aurocapilla *aw-ro-ka-PIL-a*
Aurum, gold, and *capillus*, hair, as in *Seiurus aurocapilla*, the Ovenbird, with a line of orange feathers on the top of the head that can be erected

Auroreus *aw-ROR-ee-us*
Daybreak, sunrise, as in *Phoenicurus auroreus*, the Daurian Redstart

Australis *AUS-tra-lis*
After a hypothetical southern continent, *Terra australis incognita*, as in *Acrocephalus australis*, the Australian Reed Warbler

Axillaris *ak-sil-LAR-is*
Axil, under the armpit, and *aris*, referring to, as in *Myrmotherula axillaris*, the White-flanked Antwren

Ayresii *AIRS-ee-eye*
After Thomas Ayres, British collector and naturalist, as in *Cisticola ayresii*, Wing-snapping Cisticola

Aythya *eye-THEE-a*
Greek *aithuia*, a water bird, as in *Aythya affinis*, the Lesser Scaup

---

LATIN IN ACTION

The Double-crested Cormorant has two tufts of feathers or "crests," but the Latin adjective *auritus*, long-eared, is used to describe the bird. *Auritus* can also mean attentive or nosey, as forward-facing ears on a mammal like a dog suggest. These tufts are not noticeable except during breeding season. Interestingly, the tufts tend to be black in the more southern areas of North America and get whiter as one moves northward. Alaskan birds' tufts are white. This gradual and continuous gradient of tuft color is an example of a cline, from the Greek *klinein*, meaning to lean or incline, and is one way birds recognize other birds from their population.

Phalacrocorax auritus,
Double-crested Cormorant

B

Bacchus *BAK-kus*
Roman god of wine, as in *Ardeola bacchus*, the Chinese Pond Heron, with a head and neck the color of red wine

Bachmani *BAK-man-eye*
After John Bachman, minister and naturalist who co-authored *Quadrupeds of North America* with J. J. Audubon, as in *Haematopus bachmani*, the Black Oystercatcher

Badia *ba-DEE-a*
Bay-colored, as in *Cecropis badia*, the Rufous-bellied Swallow

Badius *BA-dee-us*
Chestnut or bay colored, as in *Ploceus badius*, the Cinnamon Weaver

Baeolophus *bee-o-LO-fus*
Greek, *baio*, little, and Latin, *lophus*, crest, as in *Baeolophus bicolor*, the Tufted Titmouse

Baeopogon *bee-o-PO-gon*
Greek, *baio*, little, and *pogon*, beard, as in *Baeopogon indicator*, the Honeyguide Greenbul

Bahamensis *ba-ha-MEN-sis*
Bahamian; as in *Anas bahamensis*, the White-cheeked (or Bahama) Pintail

Baileyi *BAY-lee-eye*
After Alfred Marshall Bailey, director of the Denver Museum of Natural History, who collected the specimen, as in *Xenospiza baileyi*, Sierra Madre Sparrow

Bailloni, -ius *by-LON-eye/ee-us*
After Louis Antoine François Baillon, French naturalist and collector, as in *Baillonius* (now *Pteroglossus*) *bailloni*, the Saffron Toucanet

Bairdii *BEAR-dee-eye*
After Spencer Fullerton Baird, naturalist and second secretary of the Smithsonian Institution, as in *Calidris bairdii*, Baird's Sandpiper

Bakeri *BAY-ker-eye*
After John Randal Baker, professor at the University of Oxford, as in *Ducula bakeri*, the Vanuatu Imperial Pigeon; also George Fisher Baker, American banker and trustee of the American Museum of Natural History, as in *Sericulus bakeri*, Fire-maned Bowerbird

Baeolophus bicolor, Tufted Titmouse

Balaeniceps *bay-LEEN-ih-seps*
Balaena, whale, and *ceps*, head, as in *Balaeniceps rex*, the Shoebill

Bambusicola *bam-bus-ih-KO-la*
From *Bambuseae*, the bamboo family, and *cola*, dweller, as in *Bambusicola thoracicus*, the Chinese Bamboo Partridge

Bangsia *BANG-see-a*
After Outram Bangs, curator of mammals at the Harvard Museum of Comparative Zoology, as in *Bangsia arcaei*, the Blue-and-gold Tanager

Banksiana *bank-see-AN-a*
After Joseph Banks, English botanist and explorer, as in *Neolalage banksiana*, the Buff-bellied Monarch

Banksii *BANK-see-eye*
After Joseph Banks, English botanist and explorer, as in *Calyptorhynchus banksii*, the Red-tailed Black Cockatoo

Bannermani *BAN-ner-man-eye*
After David Armitage Bannerman, former Chairman of the British Ornithologists' Club, as in *Tauraco bannermani*, Bannerman's Turaco

Barbarus bar-BAR-us
Barba, beard, as in *Megascops barbarus*, the Bearded Screech Owl

Barbatus bar-BA-tus
Barba, beard, as in *Gypaetus barbatus*, the Bearded Vulture

Barbirostris bar-bi-ROSS-tris
Barba, beard, and *rostris*, beak, as in *Myiarchus barbirostris*, the Sad Flycatcher, the scientific name referring to rictal bristles

Barlowi BAR-lo-eye
After Charles Barlow, South African businessman, as in *Calendulauda barlowi*, Barlow's Lark

Barnardius bar-NAR-dee-us
After Edward Barnard, zoologist, botanist, and horticulturalist, as in *Barnardius zonarius*, the Australian Ringneck

Baroni BA-ron-eye
After O. T. Baron, German engineer and amateur ornithologist, as in *Cranioleuca baroni*, Baron's Spinetail

Bartletti BART-let-tye
After Abraham Bartlett, taxidermist and zoologist at the Zoological Society of London, as in *Crypturellus bartletti*, Bartlett's Tinamou

Bartramia bar-TRAM-ee-a
After William Bartram, naturalist, botanist, and explorer; his father was known as the father of American botany, as in *Bartramia longicauda*, the Upland Sandpiper

Baryphthengus bar-if-THEN-gus
Greek, *bary*, heavy, and *phthengis*, voice, as in *Baryphthengus martii*, the Rufous Motmot

Basileuterus bas-ih-LOY-ter-us
Greek, *basil-*, royal, kingly, and *euter*, music as in *Basileuterus* (now *Myiothlypis*) *fraseri*, the Gray-and-gold Warbler

Basilornis bas-ih-LORN-is
Greek, *basil-*, royal, and *ornis*, bird, as in *Basilornis celebensis*, the Sulawesi Myna

Batesi BATES-eye
After George Bates, who wrote the *Handbook of the Birds of West Africa*, as in *Apus batesi*, Bates's Swift

Bathmocercus bath-mo-SIR-kus
Greek, *bathmo*, step or degree, and Latin, *cerco*, tail, as in *Bathmocercus cerviniventris*, the Black-headed Rufous Warbler

Batis BA-tis
Malay-Polynesian, plant, as in *Batis minor*, the Eastern Black-headed Batis, after an unidentified Greek bird that fed in the brambles

Batrachostomus ba-tra-ko-STO-mus
Greek, *batracho*, frog, and *stoma*, mouth, as in *Batrachostomus moniliger*, the Sri Lanka Frogmouth

Baumanni BOW-man-nye
After Oscar Baumann, Austrian explorer and geographer, as in *Phyllastrephus baumanni*, Baumann's Olive Greenbul

Becki BECK-eye
After Rollo Beck, American bird collector, as in *Pseudobulweria becki*, Beck's Petrel

Belcheri BEL-cher-eye
After Admiral Edward Belcher, British naval officer and explorer, as in *Larus belcheri*, Belcher's Gull

Beldingi BEL-ding-eye
After Lyman Belding, American professional bird collector, as in *Geothlypis beldingi*, Belding's Yellowthroat

Bella BEL-la
Beautiful, fair, as in *Goethalsia bella*, the Pirre Hummingbird

Bellulus Bell-LU-lus
From *bellus*, pretty, as in *Margarornis bellulus*, the Beautiful Treerunner

Basilornis celebensis, Sulawesi Myna

Bendirei *ben-DEER-eye*
After Charles Emil Bendire, egg-collector, oologist, and US Army surgeon, as in *Toxostoma bendirei*, Bendire's Thrasher

Bengalensis *ben-ga-LEN-sis*
After the Bengal region shared by India and Bangladesh, as in *Bubo bengalensis*, the Indian Eagle-Owl

Berlepschi *ber-LEP-shy*
After Hans Hermann von Berlepsch, a German ornithologist, as in *Chaetocercus berlepschi*, Esmeraldas Woodstar

Berliozi *bear-lee-OZE-eye*
After Jaques Berlioz, French ornithologist, as in *Apus berliozi*, Forbes-Watson's Swift

Berthelotii *ber-te-LOT-ee-eye*
After Sabin Berthelot, French naturalist and author of the *Natural History of the Canary Islands*, as in *Anthus berthelotii*, Berthelot's Pipit

Bewickii *bee-WIK-ee-eye*
After Thomas Bewick, English naturalist and wood engraver, as in *Thryomanes bewickii*, Bewick's Wren

Bias *BY-as*
French, *biais*, slope, sideways, against the grain, as in *Bias musicus*, the Black-and-white Shrike-flycatcher

Biarmicus *Bi-ARM-i-cus*
From Biarmica, a region of Russia, as in *Falco biarmicus*, the Lanner Falcon

Biatas *by-AT-as*
Greek, forceful, mighty, as in *Biatas nigropectus*, the White-bearded Antshrike

Bicalcarata, -um, -us *by-kal-kar-AT-a/um/us*
Bi-, two, and *calcar*, spur, as in *Galloperdix bicalcarata*, the Sri Lanka Spurfowl

Bicinctus *by-SINK-tus*
Bi-, twice, and *cinctus*, encircled, surrounded, banded, as in *Treron bicinctus*, the Orange-breasted Green Pigeon, with an orange band on its chest

Bicknelli *BIK-nel-lye*
After Eugene Bicknell, American ornithologist and businessman, as in *Catharus bicknelli*, Bicknell's Thrush

---

LATIN IN ACTION

Eremophila bilopha describes Temminck's (Horned) Lark as a two-crested bird with a love of lonely places. The two crests (or tufts or horns) are actually elongated feathers that are obvious on the head of the breeding male and more subtle on the female. The bird lives in far northern Africa, extending eastward to part of the Middle East, where it inhabits rocky, semi-desert habitats. Seventeen species of birds were anointed with the common name of Temminck, after Coenraad Temminck, a Dutch zoologist who wrote a manual on the birds of Europe over the years 1815–1840 that was a standard for many years. "Lark" is from the Middle English *larke*, songbird.

Eremophila bilopha, Temminck's Lark

Bicolor *BY-ko-lor*
Bi-, two, and *color*, color, as in *Nigrita bicolor*, the Chestnut-breasted Nigrita (finch)

Bicornis *by-KOR-nis*
Bi-, two, and *cornis*, horned, as in *Buceros bicornis*, the Great Hornbill

Bidentatus, -a *by-den-TA-tus/ta*
Bi-, two, and *dentata*, teeth, as in *Lybius bidentatus*, the Double-toothed Barbet

Bifasciatus *by-fa-see-AH-tus*
Bi-, two, and *fasciatus*, banded, as in *Saxicola* (now *Campicoloides*) *bifasciatus*, the Buff-streaked Chat

Bilineata, -us *by-lin-ee-AH-ta/tus*
Bi-, two, and *linea*, line, as in *Amphispiza bilineata*, the Black-throated Sparrow

Bilopha, -us *by-LO-fa/fus*
Bi-, two, and *lophus*, crest, as in *Eremophila bilopha*, Temminck's Lark (see box)

Bimaculata, -us, *by-mak-oo-LA-ta/tus*
Bi-, two, and *maculates*, spotted, as in *Melanocorypha bimaculata*, the Bimaculated Lark

Binotata *by-no-TAT-a*
Bi-, two, and *nota*, marked, as in *Apalis binotata*, the Lowland Masked Apalis

Birostris *by-ROSS-tris*
Bi-, two, and *rostris*, beak, as in *Ocyceros birostris*, the Indian Gray Hornbill

Biscutata *bis-koo-TAT-a*
Bi-, two, and *scutum*, shield, as in *Streptoprocne biscutata*, the Biscutate Swift, in reference to the shape of the wings

Bishopi *BISH-op-eye*
After Charles Bishop, American businessman who lived in Hawaii, as in *Moho bishopi*, Bishop's Oo

Bistriatus *bis-tree-AH-tus*
Bi-, two, and *stria*, streak, as in *Burhinus bistriatus*, the Double-striped Thick-knee

Bistrigiceps *bis-TRIH-ji-seps*
Bi-, two, *striga*, furrow, and *ceps*, head, as in *Acrocephalus bistrigiceps*, the Black-browed Reed Warbler

Bitorquata, -us *by-tor-KWA-ta/tus*
Bi-, two, *torquatus*, collar, as in *Streptopelia bitorquata*, the Island Collared Dove

Bivittata, -us *bi-vit-TAT-a/us*
Bi-, two, *vitta*, stripe, band, as in *Petroica bivittata*, the Mountain Robin

Blanfordi *BLAN-for-dye*
After William Blanford, geologist and zoologist, as in *Pyrgilauda blanfordi*, Blanford's Snowfinch

Bleda *BLED-a*
After Bleda the Hun, brother of Atilla, as in *Bleda syndactylus*, the Red-tailed Bristlebill

Blythii *BLYTH-ee-eye*
After Edward Blyth, English zoologist, as in *Tragopan blythii*, Blyth's Tragopan

Blythipicus *bly-thih-PIK-us*
After Edward Blyth, English zoologist, and *picus*, woodpecker, as in *Blythipicus pyrrhotis*, the Bay Woodpecker

Bocagii *bo-KAJ-ee-eye*
After Portuguese naturalist José Vicente Barlosa du Bocage, as in *Nectarinia bocagii*, Bocage's Sunbird

Boissonneaua *bwa-son-O-a*
After Adolph Boissoneau, French ornithologist and author, as in *Boissonneaua flavescens*, the Buff-tailed Coronet

Bolbopsittacus *bol-bop-SIT-ta-kus*
Greek, *bolbo*, bulb, and *psittakos*, parrot, as in *Bolbopsittacus lunulatus*, the Guaiabero

Bolborhynchus *bol-bo-RINK-us*
Greek, *bolbo*, bulb, and Latin, *rhynchus*, bill, as in *Bolborhynchus lineola*, the Barred Parakeet

Boliviana, -us, -um *bo-liv-ee-AN-a/us/um*
After Bolivia, as in *Atilla bolivianus*, the White-eyed Atilla

Bombycilla garrulus, Bohemian Waxwing

Bollii BOL-lee-eye
After Carl Bolle, German collector and botanist, as in *Columba bollii*, Bolle's Pigeon

Bombycilla bom-bi-SIL-la
Greek, *bombyx*, silkworm, and Latin, *cilla*, hair, as in *Bombycilla garrulus*, the Bohemian Waxwing, with silky smooth plumage

Bonapartei bo-na-PAR-tye
After J. Bonaparte, American ornithologist, as in *Nothocercus bonapartei*, the Highland (or Bonaparte's) Tinamou

Bonasa bo-NA-sa
Bonasus, wild bull, as in *Bonasa umbellus*, the Ruffed Grouse; probably refers to the sound of the bird's rapid wing-beating display, known as "drumming"

Bonelli bo-NEL-lye
After Franco Bonelli, Italian ornithologist and collector, as in *Phylloscopus bonelli*, the Western Bonelli's Warbler

Boobook BOO-book
After the call of various owls in Asia and Australia, as in *Ninox boobook*, the Southern Boobook

Borbonica, -us bor-BON-ih-ka/kus
After Ile Bourbon, the former name of Ile Reunion, as in *Phedina borbonica*, the Mascarene Martin

Borealis bor-ee-AH-lis
Northern, of the north, as in *Picoides borealis*, the Red-cockaded Woodpecker, or *Phylloscopus borealis*, the Arctic Warbler

Borealoides bor-ee-a-LOID-eez
Resembling the north, as in *Phylloscopus borealoides*, the Sakhalin Leaf Warbler

Bornea BOR-nee-a
After Borneo, as in *Eos bornea*, the Red Lory

Bostrychia bo-STRICK-ee-a
Greek, *bostrych*, curl, as in *Bostrychia olivacea*, the Olive Ibis, named for its curved bill

Botaurus bo-TAW-rus
Bo, cow, and *taurus*, bull, as in *Botaurus stellaris*, the Eurasian Bittern, referring to the bird's booming call

Terpsiphone bourbonnensis, Mascarene Paradise Flycatcher

Bottae BOT-tee
After Carl-Emile Botta, French traveler and doctor, as in *Oenanthe bottae*, the Red-breasted Wheatear

Botterii bot-TARE-ee-eye
After Matteo Botteri, Yugoslavian ornithologist and collector, as in *Peucaea botterii*, Botteri's Sparrow

Boucardi boo-KARD-eye
After Adolphe Boucard, French naturalist, as in *Amazilia boucardi*, the Mangrove Hummingbird

Bougainvillei boo-gen-VIL-lye
After Louis-Antoine de Bougainville, French admiral and explorer, as in *Actenoides bougainvillei*, the Moustached Kingfisher

Bourbonnensis boor-bon-NEN-sis
After Ile Bourbon, the former name of Ile Reunion, as in *Terpsiphone bourbonnensis*, the Mascarene Paradise Flycatcher

Boweri BOW-er-eye
After Thomas Bowyer-Bower, English-born curator of ornithology in Australia, as in *Colluricincla boweri*, Bower's Shrikethrush

Boyeri BOY-er-eye
After Joseph Boyer, French sea captain and explorer, as in *Coracina boyeri*, Boyer's Cuckooshrike

Braccatus brak-KA-tus
Wearing trousers, as in *Moho braccatus*, the extinct Kauai Oo, a honeyeater, referring to its yellow-colored thighs

Avian Adaptations

Since birds arrived on the scene over 150 million years ago, they diverged into a wide variety of niches and the adaptations that provide them with the means to live successfully. In spite of their diversity, birds are probably the most homogeneous group in the animal kingdom. They are all homeothermic (warm-blooded), they all lay eggs, the vast majority show some parental care, all have feathers, and all but 40 of 10,000 species can fly.

The skeleton of birds is made to withstand the stresses of flying and landing. Many of their bones are fused, such as the caudal vertebrae, forming the pygostyle, a tail structure covered with fat and muscle and sometimes called the "pope's nose." Bones of the pelvic girdle and some bones of the arm and hand are fused. The furcula (wishbone) and uncinate (hooked) processes of the ribs stiffen the skeleton while preserving flexibility. Instead of a toothed jaw, they have a beak. To manipulate objects, they have a very flexible neck with 13–25 cervical vertebrae, compared to seven for most mammals'. Their bones are generally denser than mammalian bones, pneumatic and cross-hatched with struts, making them very strong.

Birds have large eyes with superb light-gathering power, visual acuity, and light sensitivity. They can see 180 degrees or more and keep everything in focus as their eyes are somewhat flattened. They have an enormous number of rods and cones (photoreceptive cells) in their retina. They can not only see visible light, but ultraviolet. Their lenses can change their optical properties quickly, allowing the birds to keep in focus and track objects like flying insects, and navigate through bushes and trees without collisions.

Birds' hearing is acute. Although most lack an external ear, their ear construction and the range of frequencies they can detect are similar to mammals. Owls can hear particularly well because they do have external pinna that help to capture sound, but theirs are asymmetrical so they can pinpoint the direction of the sound. Since many birds use calls or songs for courtship, identification, or territory defense, hearing is an important survival sense. In mammals the little hair cells that transmit sound from the ear to the brain die off as the animals get older, causing increasing degrees of deafness. In birds the hair cells are regenerated so they can maintain acute hearing throughout their life.

Birds use a lot of energy flying, requiring more oxygen and driving up body temperature. A very efficient respiratory and cooling system is made

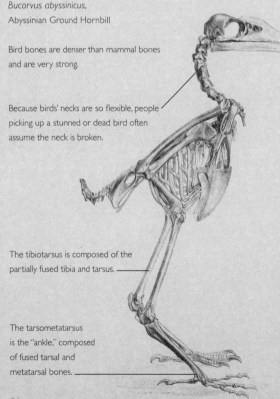

Bucorvus abyssinicus,
Abyssinian Ground Hornbill

Bird bones are denser than mammal bones and are very strong.

Because birds' necks are so flexible, people picking up a stunned or dead bird often assume the neck is broken.

The tibiotarsus is composed of the partially fused tibia and tarsus.

The tarsometatarsus is the "ankle," composed of fused tarsal and metatarsal bones.

AVIAN ADAPTATIONS

Bubo capensis,
Cape Eagle-Owl

The terminal barbs of owls' flight feathers do not hook together, producing a frayed edge for a virtually noiseless flight.

doves (Columbiformes) can suck water up into their throats, but most birds have to fill their mouths with liquid and tilt their head back. To save weight, most birds do not have a urinary bladder and have minimal water requirements. Mammals produce urea, a toxic substance that needs to be diluted before it is passed from the body via the bladder. Birds produce the insoluble uric acid which can be excreted, along with the feces (bird droppings) with very little water loss.

Flying, especially migrating long distances every year, requires all these adaptations and more, making everyday survival a tenuous business. For songbirds even a short nap on the branch of a tree requires a special adaptation. Have you ever wondered how birds can sleep without falling out of a tree? It turns out that a special tendon running from the back of the leg to the toes contracts when the bird bends its legs to perch and pulls the toes into a curled position. When it flies off, the tendon stretches and the toes uncurl. Birds are amazing, indeed.

possible by air sacs, extensions of the lungs. Although these extensions do not exchange oxygen, they provide an efficient and constant flow of air over the lungs. Birds do not have sweat glands, so air exchange across the lungs is the primary mechanism for cooling.

Since birds do not have teeth—although some have tooth-like projections on the edge of their jaws or on their palate—they cannot chew their food. They possess an expanded part of the esophagus called the crop that starts the digestion process. The partly digested food goes to the two-part stomach, the first part being the muscular crop that physically masticates the food, sometimes with the help of grit that the bird swallows. Pigeons and

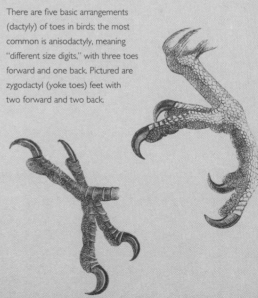

There are five basic arrangements (dactyly) of toes in birds; the most common is anisodactyly, meaning "different size digits," with three toes forward and one back. Pictured are zygodactyl (yoke toes) feet with two forward and two back.

Brachycope brak-ee-KOPE-ee
Greek, *brachy*, short, and *cope*, handle, as in *Brachycope anomala*, the Bob-tailed Weaver

Brachydactyla brak-ee-dak-TIL-a
Greek, *brachy*, short, and *dactyl*, finger or toe, as in *Certhia brachydactyla*, the Short-toed Treecreeper

Brachypteracias bra-kip-ter-ACE-ee-as
Greek, *brachy*, short, and *ptery*, wing, as in *Brachypteracias leptosomus*, the Short-legged Ground Roller

Brachypterus bra-kip-TER-us
Greek, *brachy*, short, and *ptery-*, wing, as in *Tachyeres brachypterus*, the flightless Falkland Steamer Duck

Brachyramphus bra-ki-RAM-fus
Greek, *brachy*, short, and *ramphus*, bill, as in *Brachyramphus marmoratus*, the Marbled Murrelet

Brachyrhyncos, -a bra-kee-RINK-os/a
Greek, *brachy*, short, and Latin, *rhynchus*, bill, as in *Corvus brachyrhynchos*, the American Crow

Pitta brachyura, Indian Pitta

Brachyura, -us bra-kee-OO-ra/rus
Greek, *brachy*, short, and *oura*, tail, as in *Pitta brachyura*, the Indian Pitta

Bracteatus brak-tee-AH-tus
Gold leaf, as in *Dicrurus bracteatus*, the Spangled Drongo

Bradornis brad-OR-nis
Greek, *brad*, slow, and *ornis*, bird, as in *Bradornis pallidus*, the Pale Flycatcher. Feeding on or near the ground, they are less active than other flycatchers

Bradypterus brad-ip-TER-us
Greek, *brady*, slow and *ptery*, winged, as in *Bradypterus baboecala*, the Little Rush Warbler. Unlike birds similar in appearance, this bird climbs through vegetation, skulks, and walks, rather than flies, from danger

Brandti BRANT-eye
After Johann Friedrich von Brandt, German zoologist, as in *Leucosticte brandti*, Brandt's Mountain Finch

Branickii bran-IK-ee-eye
After Heironim Graf von Branicki, Polish zoologist, as in *Leptosittaca branickii*, the Golden-plumed Parakeet

Branta BRAN-ta
Origin may be Old Norse, *brantgas*, the sheldrake, as in *Branta bernicla*, the Brant or Brent Goose

Brasiliana, -um, -us, -ensis bra-sil-ee-AN-a/um/us/bra-sil-ee-a-NEN-sis
After Brazil, as in *Cercomacra brasiliana*, the Rio de Janeiro Antbird

Brehmii BREM-ee-eye
After Alfred Brehm, German collector and zoologist, as in *Psittacella brehmii*, Brehm's Tiger Parrot

Brevicaudata bre-vi-kaw-DA-ta
Brevis, short, and *caudata*, tailed, as in *Camaroptera brevicaudata*, the Gray-backed Camaroptera

Brevipennis bre-vi-PEN-is
Brevis, short, and *pennis*, feather, as in *Acrocephalus brevipennis*, the Cape Verde Warbler

Brevipes breh-VIP-eez
Brevis, short, and *pes*, foot, as in *Accipiter brevipes*, the Levant Sparrowhawk

Brevirostris *bre-vi-ROSS-tris*
Brevis, short, and *rostris*, beak, as in *Brachyramphus brevirostris*, Kittlitz's Murrelet

Brevis *BRE-vis*
Brevis, short, as in *Bycanistes brevis*, the Silvery-cheeked Hornbill (see box)

Breweri *BREW-er-eye*
After Thomas Mayo Brewer, American ornithologist, as in *Anas breweri*, Brewer's Duck, which is actually a hybrid between the Mallard, *Anas platyrhynchos* and the Gadwall, *Anas strepera*

Brookii *BROOK-eye*
After Charles Brooke, a White Rajah of Sarawak, Malaysia, as in *Otus brookii*, Rajah Scops Owl

Browni, -ii *BROWN-eye/ee-eye*
After George Brown, English missionary to Melanesia, as in *Platycercus caledonicus brownii*, Brown's Parakeet, a subspecies of the Green Rosella

Bruijnii *BROIN-ee-eye*
After Anton Bruijn, a Dutch feather merchant, as in *Drepanornis bruijnii*, the Pale-billed Sicklebill

Brunnea *brun-NEE-a*
Brunne, brown, as in *Alcippe brunnea*, the Dusky Fulvetta

Platycercus caledonicus brownii, Brown's Parakeet (subspecies)

Brunneicapillus *brun-nee-ka-PIL-lus*
Brunne, brown, and *capilla*, cloak, as in *Aplonis brunneicapillus*, the White-eyed Starling

Brunneicauda *brun-nee-KAW-da*
Brunne, brown, and *cauda*, tail as in *Alcippe brunneicauda*, the Brown Fulvetta

Brunneiceps *BRUN-ni-seps*
Brunne, brown, and *ceps*, headed, as in *Yuhina brunneiceps*, the Taiwan Yuhina

Brunneinucha *brun-e-nee-NOO-ka*
Brunne, brown, and *nucha*, nape, as in *Arremon brunneinucha*, the Chestnut-capped Brush Finch

Brunneipectus *brun-nee-PEK-tus*
Brunne, brown, and *pectus*, neck, as in *Capito brunneipectus*, the Brown-chested Barbet

Brunneiventris *brun-nee-VEN-tris*
Brunne, brown, and *ventris*, belly, as in *Diglossa brunneiventris*, the Black-throated Flowerpiercer

---

LATIN IN ACTION

The Silvery-cheeked Hornbill's genus, *Bycanistes*, means "trumpeter," undoubtedly after its low trumpeting call. The specific epithet, *brevis*, refers to its relatively short bill, compared with other hornbills. Residents of Africa and Asia, hornbills have a long, stout, down-curved bill, with a structure unique to hornbills on top of their upper mandible. The casque varies among species and may be small, hollow, and light, or large, heavy, and reinforced with bone. The smaller casques seem to have little or no function but the larger ones may serve as resonating chambers for calls or for territorial dueling.

Brunneopygia *brun-nee-o-PI-jee-a*
Brunne, brown, and *puge*, rump, as in *Drymodes brunneopygia*, the Southern Scrub Robin

Brunneus *BRUN-nee-us*
Brunne, brown, as in *Pycnonotus brunneus*, the Asian Red-eyed Bulbul

Brunnicephalus *brun-ni-se-FAL-us*
Brunne, brown, and *cephala*, head, as in *Choicocephalus brunnicephalus*, the Brown-headed Gull

Brunniceps *BRUN-ni-seps*
Brunne, brown, and *ceps*, headed, as in *Myioborus brunniceps*, the Brown-capped Whitestart

Brunnifrons *BRUN-ni-fronz*
Brunne, brown, and *frons*, forehead, as in *Cettia brunnifrons*, the Gray-sided Bush Warbler

Bubo scandiacus, Snowy Owl

Bubalornis *boo-ba-LOR-nis*
Greek, *bubal*, buffalo, and *ornis*, bird, as in *Bubalornis niger*, the Red-billed Buffalo Weaver, which often associates with cattle

Bubo *BOO-bo*
Swelling, as in *Bubo scandiacus*, the Snowy Owl, or *Bubo bubo*, the Eurasian Eagle-Owl, although the name *Bubo* probably came from the owl's deep and resonant calls

Bucco *BOO-ko*
Bucca, mouth, as in *Bucco tamatia*, the Spotted Puffbird, with a distinctly large bill and mouth

Bucephala *boo-se-FAL-a*
Greek, *bous*, ox, and Latin, *cephala*, head, as in *Bucephala clangula*, the Common Goldeneye. The shape of the head reminded the namer of an ox's head

Bucorvus *boo-KOR-vus*
Greek, *bu*, ox, and Latin, *corvus*, raven, as in *Bucorvus abyssinicus*, the Abyssinian Ground Hornbill. *Bu* can also refer to the large size of an ox and this is a large bird

Bulleri *BUL-ler-eye*
After Walter Lawry Buller, a New Zealand lawyer, naturalist and ornithologist, as in *Puffinus bulleri*, Buller's Shearwater

Bullocki *BUL-lok-eye*
After William Bullock, an amateur American ornithologist with a traveling museum, as in *Icterus bullocki*, Bullock's Oriole

Burchelli *BUR-chel-lye*
After William John Burchell, English explorer and naturalist, as in *Pterocles burchelli*, Burchell's Sandgrouse

Burhinus *bur-HINE-nus*
Greek, *bous*, ox, and *rhin*, nose or beak, as in *Burhinus capensis*, the Spotted Thick-knee

Buteo *BOO-tee-o*
Etymology unclear but a kind of hawk, as in *Buteo buteo*, the Common Buzzard

Buteogallus *boo-tee-o-GAL-lus*
Buteo, hawk, and *gallus*, cock or rooster, as in *Buteogallus anthracinus*, the Common Black Hawk

Buthraupis *boo-THRAW-pis*
Greek, *bu*, ox, and *thraupis*, tanager, as in *Buthraupis montana*, the Hooded Mountain Tanager

C

Cabanisi *ka-BAN-nis-eye*
After Jean Louis Cabanis, German founder and editor of *Journal für Ornithologie*, as in *Emberiza cabanisi*, Cabanis's Bunting

Caboti *CAB-ot-i*
After Samuel Cabot, American physician and ornithologist, as in *Tragopan caboti*, Cabot's Tragopan

Cacatua *ka-ka-TOO-a*
Dutch, *kakatoe*, or Malay, *kokatua*, cockatoo, as in *Cacatua sulphurea*, the Yellow-crested Cockatoo

Cachinnans *ka-CHIN-nans*
Laughing, as in *Herpetotheres cachinnans*, the Laughing Falcon or Snake Hawk, after its loud call that resembles laughter

Cacomantis *ka-ko-MAN-tis*
Greek, *caco-*, bad, ill omen, and *mantis*, a seer or prophet, as in *Cacomantis merulinus*, the Plaintive Cuckoo; the cuckoo was thought to be able to foretell the future

Cactorum *kak-TOE-rum*
Greek, *kaktos*, cactus, as in *Melanerpes cactorum*, the White-fronted Woodpecker, which inhabits environments with cacti

Caerulea *see-ROO-la*
Referring to sky, or sea, or blue, as in *Passerina caerulea*, the Blue Grosbeak

Caerulatus *see-roo-LA-tus*
Referring to sky, or sea, or blue, as in *Cyornis caerulatus*, the Sunda Blue Flycatcher

Caeruleirostris *see-roo-lee-eye-ROSS-tris*
Caerul, blue, and *rostris*, bill or beak, as in *Loxops caeruleirostris*, the Akekee, a honeycreeper with a blue bill

Caeruleogrisea *see-roo-lee-o-GRISS-ee-a*
Caerul, blue, and *grisea*, gray, as in *Coracina caeruleogrisea*, the Stout-billed Cuckooshrike

Caerulescens *see-roo-LES-sens*
Referring to sky, or sea, or blue, as in *Chen caerulescens*, the Snow or Blue Goose, because it has a blue morph

Passerina caerulea, Blue Grosbeak

Caeruleus *see-ROO-lee-us*
Sky blue, as in *Cyanocorax caeruleus*, the Azure Jay

Caeruleogularis *see-roo-le-o-goo-LAR-is*
Caerul, blue, and *gularis*, throat, as in *Aulacorhynchus caeruleogularis*, the Blue-throated Toucanet

Caesia, -us *SEE-zee-a/us*
Pertaining to Caesar's eyes, which were gray or gray-blue, as in *Thamnomanes caesius*, the bluish-gray Cinereous Antshrike

Cafer *KAY-fer*
After South Africa, as in *Pycnonotus cafer*, the Red-vented Bulbul. The bird was mistakenly first named after South Africa

Cahow *KA-how*
Imitative of the bird's call, as in *Pterodroma cahow*, the Bermuda Petrel, known in Bermuda as the Cahow

Cairina *ky-REE-na*
After Cairo, Egypt, as origin, as in *Cairina moschata*, the Muscovy Duck, which is actually from South America

Calamanthus *ka-lam-AN-thus*
Greek, *kalame*, a stalk of grain, and *anthus*, flower, as in *Calamanthus campestris*, the Rufous Fieldwren

Calamonastes *kal-a-mo-NAS-teez*
Greek, *kalame*, a stalk of grain, and *astes*, singer, as in *Calamonastes simplex*, the Gray Wren-Warbler

Calamospiza *kal-a-mo-SPY-za*
Greek, *kalame*, a stalk of grain, and *spiza*, finch, as in *Calamospiza melanocorys*, the Lark Bunting

Calcarius *kal-KAR-ee-us*
Calx, of lime, or limestone, or heel, spur, as in *Calcarius lapponicus*, the Lapland Longspur or Bunting, which has a long rear toe

Calendula *ka-len-DOO-la*
Calendae, little calendar or little clock, as in *Regulus calendula*, the Ruby-crowned Kinglet, perhaps having to do with the timing of its appearance during migration

Caledonica, -us *kal-ih-DON-ih-ka/us*
After New Caledonia, as in *Coracina caledonica*, the South Melanesian Cuckooshrike and *Nycticorax caledonicus*, the Nankeen Night Heron

Calicalicus *Cal-i-CAL-i-cus*
Derives from the local Madagascan name, Calicalac, of *Calicalicus madagascariensis*, the Red-tailed Vanga

Californianus, Californica *kal-ih-for-nee-AN-us/kal-ih-FOR-ni-ka*
After California, as in *Geococcyx californianus*, the Greater Roadrunner, and *Aphelocoma californica*, the California Scrub-Jay

Callacanthis *kal-la-KAN-this*
Greek, *kallos*, beautiful, and *acanthis*, a (gold) finch, as in *Callacanthis burtoni*, the Spectacled Finch

Calliope *kal-LY-o-pee*
Greek, *kallos*, beautiful, and *ops*, voice, as in *Luscinia calliope*, the Siberian Rubythroat

Callipepla *kal-li-PEP-la*
Greek, *kallos*, beautiful, and *pepla*, robe, as in *Callipepla californica*, the California Quail

Calliphlox *KAL-li-flox*
Greek, *kallos*, beautiful, and *phlox*, a flower, as in *Calliphlox amethystina*, the Amethyst Woodstar

Callocephalon *kal-lo-se-FAL-on*
Greek, *kallos*, beautiful, and Latin, *cephala*, head, as in *Callocephalon fimbriatum*, the Gang-gang Cockatoo; Gang-gang comes from an Aboriginal language

Callonetta *kal-lo-NET-ta*
Greek, *kallos*, beautiful, and *netta*, duck, as in *Callonetta leucophrys*, the Ringed Teal

Calochaetes *kal-o-KEE-teez*
Greek, *kallos*, beautiful, and *chaete*, long flowing hair, mane, as in *Calochaetes coccineus*, the Vermilion Tanager; the feathers of the nape and wing coverts resemble a mane

Calocitta *kal-o-SIT-ta*
Greek, *kallos*, beautiful, and Latin, *citta*, magpie, jay, as in *Calocitta formosa*, the White-throated Magpie-Jay

Calonectris *kal-o-NEK-tris*
Greek, *kallos*, beautiful, and *nectris*, swimmer, as in *Calonectris leucomelas*, the Streaked Shearwater

Caloperdix *kal-o-PER-diks*
Greek, *kallos*, beautiful, and *perdix*, partridge, as in *Caloperdix oculeus*, the Ferruginous Partridge

Calopterus *kal-OP-ter-us*
Greek, *kallos*, beautiful, and *ptery*, wing, as in *Mecocerculus calopterus*, the Rufous-winged Tyrannulet

Calothorax *kal-o-THOR-aks*
Greek, *kallos*, beautiful, and *thorax*, breast or chest, as in *Calothorax pulcher*, the Beautiful Sheartail

Calvus *KAL-vus*
Bald, as in *Sarcops calvus*, the Coleto (in the starling family)

Regulus calendula, Ruby-crowned Kinglet

Calliphlox amethystina,
Amethyst Woodstar

Calypte *ka-LIP-tee*
Greek, *calypto*, veiled, mantled, or hidden, as in *Calypte anna*, Anna's Hummingbird; may have to do with the head being covered in iridescent red feathers

Calyptocichla *kal-ip-toe-SIK-la*
Greek, *calypto*, hidden, and *cichla*, thrush, as in *Calyptocichla serinus*, the Golden Greenbul, thrush-like in appearance

Calyptomena *kal-ip-toe-MEN-a*
Greek, *calypto*, hidden, and *mena*, moon, as in *Calyptomena viridis*, the Green Broadbill, referring to the bill being mostly hidden by tufts of feathers

Calyptophilus *ka-lip-toe-FIL-us*
Greek, *calypto*, hidden, and *phila*, love, as in *Calyptophilus tertius*, the Western Chat-Tanager, a secretive bird preferring dense undergrowth on the forest floor

Calyptorhynchus *ka-lip-tow-RINK-us*
Greek, *calypto*, hidden, and Latin, *rhynchus*, bill, as in *Calyptorhynchus banksii*, the Red-tailed Black Cockatoo, with a partially hidden beak

Camaroptera *kam-a-ROP-ter-a*
Greek, *kamara*, arch, and *ptery*, wing, as in *Camaroptera brachyura*, the Green-backed Camaroptera; the name may have to do with the way the bird holds its wings slightly away from the body

---

LATIN IN ACTION

The familiar Common Ostrich, *Struthio camelus*, is described by its scientific name, as the "camel sparrow," camel for its mammalian neighbors, but the description as a sparrow does not quite fit. At nearly 10 feet (3 meters) tall and 330 pounds (150 kilograms), they are the largest living birds in the world. Eight species of ostriches have become extinct since their evolution about 40 million years ago. The Common Ostrich is distantly related to the other groups of large flightless birds such as emus, cassowaries, rheas, and kiwis. These birds are tied together in a group called ratites, birds without a keel on their sternum to anchor flight muscles.

Cambodiana *kam-bo-dee-AN-a*
After Cambodia, as in *Arborophila cambodiana*, the Chestnut-headed Partridge

Camelus *kam-EL-us*
Camel, dromedary, as in *Struthio camelus*, the Common Ostrich (see box), the reference to camel alluding to its dry habitat

Camerunensis *ka-mee-roo-NEN-sis*
After Cameroon, as in *Vidua camerunensis*, the Cameroon Indigobird

Campanisoma *kam-pa-ni-SO-ma*
Campan, bell, and Greek, *soma*, body, as in *Myrmothera campanisona*, the Thrush-like Antpitta, whose very short tail gives its body a bell-like shape

Campephaga *kam-pee-FAY-ga*
Greek, *camp*, caterpillar, and *phagein*, eat, as in *Campephaga flava*, the Black Cuckooshrike

Campephilus *kam-pe-FIL-us*
Camp, of the fields, and Greek, *philos*, love, as in *Campephilus pollens*, the Powerful Woodpecker

Campestris *kam-PESS-tris*
Campestris, deity of the fields, country goddess, as in *Calamanthus campestris*, the Rufous Fieldwren

Camptorhynchus *kamp-tow-RIN-kus*
Greek, *campto*, curve, and Latin, *rhynchus*, bill, as in *Camptorhynchus labradorius*, the extinct Labrador Duck which had a slightly upcurved bill

Camptostoma *kamp-to-STO-ma*
Greek, *campto*, curve, and *stoma*, mouth, as in *Camptostoma imberbe*, the Northern Beardless Tyrannulet, with an arched culmen (upper ridge of beak)

Campylopterus *kam-pee-LOP-ter-us*
Greek, *campo*, bending, and *pteryx*, wing, as in *Campylopterus pampa*, the Wedge-tailed Sabrewing

Campylorhynchus *kam-pee-lo-RINK-us*
Greek, *campo*, bending, and Latin, *rhynchus*, bill, as in *Campylorhynchus zonatus*, the Band-backed Wren, with a downcurved bill

Camurus *ka-MOO-rus*
Curved or arched, as in *Tockus camurus*, the Red-billed Dwarf Hornbill with a curved bill

Canadensis *ka-na-DEN-sis*
After Canada or the far north, as in *Grus canadensis*, the Sandhill Crane

Cancellata *kan-sel-LA-ta*
Latticework, as in *Prosobonia cancellata*, the Kiritimati Sandpiper, perhaps describing the variable mottling and streaking on the back and breast

Candei *KAN-dee-eye*
White, brilliant, as in *Manacus candei*, the White-collared Manakin

Candida *kan-DEE-da*
Bright, clear, as in *Amazilia candida*, the White-bellied Emerald

Canens *KAN-enz*
In Roman Myth, Canens was the personification of song, as in *Arremonops conirostris*, the Black-striped Sparrow, with a conical bill and pleasant song

Canicapillus *kan-ih-ka-PIL-lus*
Canus, gray, and *capilla*, hair, as in *Dendrocopos canicapillus*, the Gray-capped Pygmy Woodpecker

Caniceps *KAN-ih-seps*
Canus, gray, and *ceps*, head, as in *Psittacula caniceps*, the Nicobar Parakeet

Canicollis *kan-ih-KOL-lis*
Canus, gray, and *collis*, collar, as in *Ortalis canicollis*, the Chaco Chachalaca

Canicularis *kan-ih-koo-LAR-is*
Canus, gray, and *cularis*, partial circle or half moon, as in *Eupsittula canicularis*, the Orange-fronted Parakeet (or Half-moon Conure)

Canifrons *KAN-ih-fronz*
Canus, gray, and *frons*, forehead, as in *Spizixos canifrons*, the Crested Finchbill

Canigularis *kan-ih-goo-LAR-is*
Canus, gray, and *gularis*, throat, as in *Chlorospingus canigularis*, the Ashy-throated Bush Tanager

Canorus *kan-OR-us*
Pertaining to melody or tune, as in *Cuculus canorus*, the Common Cuckoo. Its song doesn't have much of a tune but is well-known

Cantans *KAN-tanz*
Singing, of song, as in *Cisticola cantans*, the Singing Cisticola

Canus *KAN-us*
White or gray, as in *Agapornis canus*, the Gray-headed Lovebird

Canutus *kan-OO-tus*
Possibly from Denmark's King Canute, as in *Calidris canutus*, the Red Knot

Capense, -is *ka-PEN-see/sis*
Of the cape, as in *Zonotrichia capensis*, Rufous-collared Sparrow, referring to southern capes such as Cape Horn and Cape of Good Hope, as in *Daption capense*, the Cape Petrel

Capitalis *kap-ih-TAL-is*
Referring to the head, as in *Grallaria capitalis*, the Bay Antpitta, perhaps because of the top of the head being darker

Capitata, -us *kap-ih-TA-ta/tus*
Capit-, headed, as in *Paroaria capitata*, Yellow-billed Cardinal, with a distinctive red head

Capito *ka-PEE-to*
Capito, large head, as in *Tregellasia capito*, the Pale-yellow Robin, referring to the large-appearing head

Caprimulgus *ka-pri-MUL-gus*
Capri-, goat, and *mulg*, to milk, as in *Caprimulgus europaeus*, the European Nightjar; the scientific name comes from the old idea that these large-mouthed birds suckled on goats

Caracara *ka-ra-KA-ra*
Native Indian name for bird after its call, as in *Caracara cheriway*, the Northern Crested Caracara

Cardinalis cardinalis, Northern Cardinal

Carbo KAR-bo
Glowing coal, charcoal, as in *Cepphus carbo*, the Spectacled Guillemot, in reference to its dark gray to nearly black plumage

Cardinalis kar-di-NAL-is
Principal or chief, as in *Cardinalis cardinalis*, the Northern Cardinal

Carduelis kar-doo-EL-is
Carduelis, goldfinch or thistlefinch, as in *Carduelis carduelis*, the European Goldfinch

Carolinae kar-o-LIN-ee
After Carolina, as in *Horornis carolinae*, the Tanimbar Bush Warbler, after the Caroline Islands in the South Pacific

Carolinensis kaa-ro-li-NEN-sis
After Carolina, as in *Sitta carolinensis*, the White-breasted Nuthatch

Carolinus kar-o-LINE-us
After Carolina, as in *Euphagus carolinus*, the Rusty Blackbird

Sitta carolinensis, White-breasted Nuthatch

Carpococcyx kar-po-KOK-siks
Greek, *carpo*, fruit, and *coccyx*, cuckoo, as in *Carpococcyx viridis*, the Sumatran Ground Cuckoo

Carunculata ka-run-koo-LA-ta
Caruncul, a bit of flesh, as in *Paradigalla carunculata*, the Long-tailed Paradigalla; refers to the bird's colorful facial wattles

Carunculatus kar-un-koo-LAT-us
Caruncul, a bit of flesh, as in *Grus carunculata*, the Wattled Crane

Caryothraustes kar-ee-o-THRAWS-teez
Greek, *caryo*, a nut, and *thrausted*, crack, as in *Caryothraustes canadensis*, the Yellow-green Grosbeak, with a poweful beak for cracking nuts

Cassini KAS-sin-eye
After John Cassin, American ornithologist and first serious, accomplished bird taxonomist, as in *Vireo cassinii*, Cassin's Vireo

Castanea, -us kas-TAN-ee-a/us
Chestnut-brown colored, as in *Anas castanea*, the Chestnut Teal and *Myophonus castaneus*, the Brown-winged Whistling Thrush

Castaneiceps kas-tan-ee-EYE-seps
Castanea, chestnut-brown colored, and *ceps*, head, as in *Ploceus castaneiceps*, the Taveta Weaver

Castaneicollis kas-tan-ee-eye-KOL-lis
Castanea, chestnut-brown colored, and *collis*, collared, as in *Pternistis castaneicollis*, the Chestnut-naped Francolin

Castaneiventris kas-tan-ee-eye-VEN-tris
Castanea, chestnut-brown colored, and *ventris*, belly, as in *Monarcha castaneiventris*, the Chestnut-bellied Monarch

Castaneocapilla kas-tan-ee-o-ka-PIL-la
Castanea, chestnut-brown colored, and *capilla*, hair, as in *Myioborus castaneocapilla*, the Tepui Whitestart

Castaneocoronata kas-tan-ee-o-ko-ro-NA-ta
Castanea, chestnut-brown colored, and *coronatus*, crowned, as in *Cettia castaneocoronata*, the Chestnut-headed Tesia (Polish for loved by god)

Castanotis kas-tan-O-tis
Castanea, chestnut-brown colored, and *oto*, ear, as in *Pteroglossus castanotis*, the Chestnut-eared Aracari

Castanotus kas-tan-O-tus
Castanea, chestnut-brown colored, and *noto*, back, as in *Turnix castanotus*, the Chestnut-backed Buttonquail

Cathartes ka-THAR-teez
Greek, *katharos*, clean, pure, as in purifier or purger, as in *Cathartes aura*, the Turkey Vulture, which scavenges, thereby clearing away dead animals

Catharus ka-THAR-us
Greek, *kathartes*, cleanser, as in *Catharus gracilirostris*, the Black-billed Nightingale-Thrush, probably referring to the song of the bird

Caudata, -us kaw-DA-ta/tus
Cauda, tail, as in *Turdoides caudata*, the Common Babbler

Caudifasciatus kaw-di-fas-se-AH-tus
Cauda, tail, and *fasciatus*, banded, as in *Tyrannus caudifasciatus*, the Loggerhead Kingbird

Cauta KAW-ta
To search, as in *Thalassarche cauta*, the Shy Albatross

Cayana kye-EN-a
After Cayenne, a city in French Guiana, as in *Cotinga cayana*, the Spangled Cotinga, common name from the Tupi language of Brazil

Cayanensis kye-a-NEN-sis
After Cayenne, a city in French Guinea, as in *Icterus cayanensis*, the Epaulet Oriole

Cecropis se-KROP-is
After Kekrops (Cecrops), an early king of Attika and founder of Athens, depicted as a man with a serpent's tail in place of legs, as in *Cecropis cucullata*, the Greater Striped Swallow, with long tail feathers

Celata se-LA-ta
Hidden, as in *Leiothlypis celata*, the Orange-crowned Warbler, referring to the usually hidden orange crown

Celebensis sel-a-BEN-sis
Refers to the Celebes Islands, now known as Sulawesi, as in *Basilornis celebensis*, the Sulawesi Myna

Centrocercus sen-tro-SIR-kus
Greek, *kentron*, spur, and *kerko*, point, as in *Centrocercus urophasianus*, the Sage Grouse

Cephalopterus ornatus, Amazonian Umbrellabird

Centropus sen-TRO-pus
Greek, *kentron*, point, and *pous*, foot, as in *Centropus burchelli*, Burchell's Coucal, referring to the long hind toe. Coucal from the French, perhaps from *couc(ou)*, cuckoo, and *al(ouette)*, lark

Cephalopterus se-fal-OP-ter-us
Cephala, head, and Greek, *pteryx*, wing, as in *Cephalopterus ornatus*, the Amazonian Umbrellabird

Cephalopyrus se-fal-o-PY-rus
Cephala, head, and Greek, *pyro*, flame (colored), as in *Cephalopyrus flammiceps*, the Fire-capped Tit

Centurus sen-TOO-rus
Greek, *kentron*, point, and *oura*, tail, as in *Centurus* (now *Melanerpes*) *carolinus*, the Red-bellied Woodpecker, referring to the pointed tail of all woodpeckers (red-bellied is an odd name as there is merely a wash of pink on the abdomen)

Cepphus SEP-fus
Greek, *kepphos*, meaning seabird, as in *Cepphus columba*, the Pigeon Guillemot

Cercococcyx ser-ko-KOK-siks
Greek, *cerco*, tail, and *coccyx*, cuckoo, as in *Cercococcyx olivinus*, the Olive Long-tailed Cuckoo

Cercomacra sir-ko-MAK-ra
Greek, *cerco*, tail, and *macro*, large, long, as in *Cercomacra serva*, the Black Antbird

David Lambert Lack
(1910–1973)

David Lambert Lack perhaps had more influence on field ornithology than any other ornithologist. Whilst still an amateur, Lack became the leading British ornithologist of his time and a respected evolutionary biologist, ecologist, and population biologist. Among his many achievements, he was director of the Edward Grey Institute of Field Ornithology at Oxford University, fellow of the Royal Society, and president of both the International Ornithological Congress and the British Ecological Society.

The son of a well-known and prosperous London surgeon, David Lack (born in 1910) lived a sumptuous life in a house with seven servants. He began learning about birds at an early age, compiling his first life list at the age of nine and identifying 100 species by the age of 15. Before he even entered college he published his first scientific paper. He attended Cambridge University, where he was elected president of the Cambridge Ornithological Club and began a friendship with Julian Huxley, an influential British evolutionist and proponent of natural selection.

From 1933 to 1940 he taught at Dartington Hall, a progressive private school, except for a year's leave in 1938 when he spent time studying the birds of the Galapagos. During World War II he served with the Army Operational Research Group, helping to develop radar. This experience was valuable later, enabling him to use radar in his studies of bird

Erithacus rubecula,
European Robin

The European Robin, pictured here, and the distantly related American Robin are commonly seen in literature, in folklore, and as holiday icons.

migration. In 1945 he became a professional ornithologist and served as director of the Edward Grey Institute of Field Ornithology, Oxford, until his death.

Lack's first substantial work, the popular *Life of the Robin* (1943), has informative and entertaining chapters on the life history of the bird, including song, flight, territory, and age, as a result of using color banding (or ringing) and other simple techniques over a four-year period. He was one of the originators of avian life history studies in Britain and influenced ornithologists around the world. He also had some ideas that were novel for the time. Lack debunked the idea that robins sing because they are happy or because they are trying to attract a female. He concluded that the song is to warn off rivals as

"Like many other naturalists, I was often as a boy exalted by natural beauty but this happened less often as I grew older, though when it came it was more intense."

David Lambert Lack

part of maintaining a territory. He also emphasized the idea that bird clutches will only be as large as the food supply would support but that they would be as large as possible.

Emanating from his meticulous field studies in the Galapagos and his measurement of the beaks of 8,000 museum skins at the Museum of Natural History in New York, *Darwin's Finches* was perhaps his most famous and influential work. It provides a fascinating account of the 14 specialized species of finch that have evolved from an original stock of seed-eating finches. This book became a classic of ornithology. Before the publication of this work, biology books of the time never mentioned the finches of the Galapagos. Now biology, zoology, ecology, and evolutionary-themed books all do, and because of Lack's book, the birds are commonly called "Darwin's Finches."

Such fieldwork inevitably led Lack to the consideration of more theoretical questions. In particular, he studied the factors controlling numbers in natural populations and concluded that such factors act more severely when numbers are high than when they are low. The irregularities of population fluctuation suggested to Lack that the control mechanisms must be very complex. He discusses these ideas in *Natural Regulation of Animal Numbers* (1954) and *Population Studies of Birds* (1966). His theory was variously interpreted by professionals like Richard Dawkins, who claimed that it supported the theory of the "selfish gene."

Lack's ideas on speciation, ecological isolation, group selection, migration, and the evolution of reproductive strategies are best summarized in his two most influential books, *The Natural Regulation of Animal Numbers* (1954) and *Ecological Adaptations for Breeding in Birds* (1968). His ideas ushered in a new field of thought and he is often considered the "father of evolutionary ecology."

When Lack died in 1973, he was working on a study of bird populations on the islands of the West Indies, a return to an earlier interest in island avifaunas. Although it still needed editing to prepare it for publication, his research on the topic was completed before his death at age 63.

The Galapagos Finches are not distinguished by their plumage, as it is mainly brown to black, but by their differing beak sizes which allow species to share a habitat.

Geospiza magnirostris, Large Ground Finch

The Large Ground Finch, largest of Darwin's Finches, specializes in eating large, hard seeds off the ground.

Cercomela sir-ko-MEL-a
Greek, *cerco*, tail, and *melas*, black, as in *Cercomela* (now *Oenanthe*) *familiaris*, the Familiar Chat

Cercotricha sir-ko-TRICK-a
Greek, *cerco*, tail, and *trikhas*, thrush, as in *Cercotricha* (now *Erythropygia*) *signata*, the Brown Scrub Robin, referring to the typical thrush's tail

Certhia SIR-thee-a
Greek, *kethios*, a tree creeper, as in *Certhia brachydactyla*, the Short-toed Treecreeper

Ceryle sir-IL-ee
Greek, *kerulos*, a sea bird, as in *Ceryle rudis*, the Pied Kingfisher, more likely to be found along rivers than the sea

Chaetocercus kee-to-SIR-kus
Greek, *chaeto*, spine or hair, and *cerco*, tail, as in *Chaetocercus mulsant*, the White-bellied Woodstar, with a double-pointed tail

Chaetoptila kee-top-TIL-a
Greek, *chaeto*, spine or hair, and *ptilon*, feather, as in *Chaetoptila angustipluma*, the extinct Kioea, distinguished by the bristle-like feathers of the head and neck

Chaetorhynchus kee-tow-RINK-us
Greek, *chaeto*, spine or hair, and Latin, *rhynchus*, bill, as in *Chaetorhynchus papuensis*, the Pygmy Drongo

Chaetura kee-TOO-ra
Greek, *chaeto*, spine or hair, and *oura*, tail, as in *Chaetura fumosa*, the Costa Rican Swift; the Swift's tail is typically very short with stiff feather shafts that allow it to perch vertically on cliff walls

Progne chalybea,
Gray-breasted Martin

Chalcomelas kal-ko-MEL-as
Greek, *chalco*, copper, and *melas*, black or dark, as in *Cinnyris chalcomelas*, the Violet-breasted Sunbird

Chalcomitra kal-ko-MIT-ra
Greek, *chalco*, copper, and Latin *mitra*, cap, as in *Chalcomitra amethystina*, the Amethyst Sunbird

Chalcopsitta kal-kop-SIT-ta
Greek, *chalco*, copper, and Latin *psitta*, parrot, as in *Chalcopsitta atra*, the Black Lory, with gold-bronze colored underwing and tail

Chalybea ka-lib-BEE-a
Steel, as in *Progne chalybea*, the Gray-breasted Martin, referring to the blue-gray color of the back

Chamaea ka-MEE-a
Greek, on the ground, low, as in *Chamaea fasciata*, the Wrentit, that spends most of its time in the brush

Chapmani CHAP-man-eye
After Frank Chapman, curator of ornithology for the American Museum of Natural History, New York, as in *Chaetura chapmani*, Chapman's Swift

Charadrius kar-A-dree-us
Plover, as in *Charadrius vociferous*, the Killdeer

Chasiempis kas-ee-EM-pis
Greek, *chasma*, a gap, and *empis*, a gnat, as in *Chasiempis ibidis*, the Oahu Elepaio, a flycatcher endemic to Hawaii; refers to the bird's insect-catching lifestyle

Chelictinia kel-ik-TIN-ee-a
Greek, *chelidon*, swallow, and *ictin*, a kite, as in *Chelictinia riocourii*, the Scissor-tailed Kite; birds called kites were named after the toy

Chelidoptera kel-ih-DOP-ter-a
Greek, *chelidon*, swallow, and *ptery*, wing, as in *Chelidoptera tenebrosa*, the Swallow-winged Puffbird

Chen KEN
Greek, goose, as in *Chen rossii*, Ross's Goose

Chenonetta ken-o-NET-ta
Greek, *chen*, goose, and *netta*, duck, as in *Chenonetta jubata*, the Maned Duck

Accipiter chilensis, Chilean Hawk

Chlorocercus klo-ro-SIR-kus
Greek, *chloro-*, green, and *cerco*, tail, as in *Lorius chlorocercus*, the Yellow-bibbed Lory

Chloroceryle klo-ro-se-RIL-ee
Greek, *chloro-*, green, and *ceryle*, kingfisher, as in *Chloroceryle amazona*, the Amazon Kingfisher; Kingfisher comes from "king of the fishers"

Chlorophonia klo-ro-FONE-ee-a
Greek, *chloro-*, green, and *phono-*, voice, as in *Chlorophonia cyanea*, the Blue-naped Chlorophonia

Chloropus klor-O-pus
Greek, *chloro-*, green, and *pous*, foot, as in *Gallinula chloropus*, the Common Moorhen

Chordeiles kor-de-IL-eez
A stringed instrument, dance, moving around (unclear), as in *Chordeiles minor*, the Common Nighthawk; name may derive from the bird's circling the sky catching insects in the evening

Chrysia KRIS-ee-a
Chrys, gold, as in *Geotrygon chrysia*, the Key West Quail-Dove which is rusty-cinnamon above with an overlay of iridescent colors, imparting a goldish sheen

Ciconia si-KO-nee-a
Stork, as in *Ciconia ciconia*, the White Stork (see box)

Childonias kil-DON-ee-as
Greek, *kheldonias*, referring to a swallow, probably because the bird resembles a large swallow, as in *Childonias hybrida*, the Whiskered Tern

Chilensis chi-LEN-sis
After Chile, as in *Accipiter chilensis*, the Chilean Hawk

Chimaera ky-MEE-ra
After the ancient Greek mythical beast made of parts of different animals, as in *Uratelornis chimaera*, the Long-tailed Ground Roller, that looks as if it was made of different birds

Chinensis chy-NEN-sis
After China, where it was first described, as in *Oriolus chinensis*, the Black-naped Oriole

Chloephaga klo-ee-FAY-ga
Chloe, yellow or yellowish and Greek, *phagin*, to eat, as in *Chloephaga hybrida*, the Kelp Goose, which eats green algae and other green plants

Chlorocephalus klo-ro-se-FAL-us
Greek, *chloro-*, green, and Latin, *cephala*, head, as in *Oriolus chlorocephalus*, the Green-headed Oriole

---

LATIN IN ACTION

It is not unusual for the genus and species to be identical, as in *Ciconia ciconia*. This may not seem very descriptive but since the White Stork is so familiar, it works. "Stork" may derive from the Old English *storc*, stiff or strong, describing the bird's upright posture. Common across Europe, associated with human dwellings and often building large nests on them, these birds feature in many myths and legends. There are numerous explanations for the idea that storks deliver babies, but the best may be that it was used by parents to tell children about the new baby in the house without embarrassment.

Cinclus SINK-lus
Greek, *kinklos*, thrush that lives near water, as in *Cinclus cinclus*, the White-throated Dipper, which feeds and nests streamside

Cincta, -us SINK-ta/tus
Cingere, surround, encircle, as in *Riparia cincta*, the Banded Martin, with a band of brown encircling its breast

Cinereicauda sin-air-ee-eye-KOW-da
Cinus, ashes, and *cauda*, tail, as in *Lampornis cinereicauda*, the Gray-tailed Mountaingem

Cinereiceps sin-air-ee-EYE-seps
Cinus, ashes, and *ceps*, headed, as in *Malacocincla cinereiceps*, the Ashy-headed Babbler

Cinereus sin-AIR-ee-us
Cinus, ashes, ash-colored, as in *Xolmis cinereus*, the Gray Monjita

Cinnyris SIN-ni-ris
From the Greek Hesychius of Alexandria, who called some unknown bird *kinnuris*, as in *Cinnyris coquerellii*, the Mayotte Sunbird

Circus SIR-kus
Circus, race course, as in *Circus cyaneus*, the Hen Harrier, which hunts in a more or less circular course

Cirrhata sir-HA-ta
Curly-headed, as in *Fratercula cirrhata*, the Tufted Puffin, known for the yellow tufts extending back from its eyes

Cisticola sis-ti-KO-la
Cista, a wooden basket, and *colo*, dwell, as in *Cisticola natalensis*, the Croaking Cisticola, whose nest is in the shape of a ball or basket

Cistothorus sis-tow-THOR-us
Greek, *kistos*, shrub, and *thorus*, a bed, as in *Cistothorus palustris*, the Marsh Wren, which hides its nest in shrubs

Citrina si-TRY-na
Citrus or lemon tree, as in *Setophaga citrina*, the Hooded Warbler, with a lemon-yellow face

Clangula klang-GOO-la
Clangere, to resound, as in *Clangula hyemalis*, the Long-tailed Duck, after the bird's distinctive call

Clypeata kli-pee-AH-ta
Clypeum, shield, as in *Anas clypeata*, the Northern Shoveler, referring to its spoon-shaped bill

Coccyzus KOK-si-zus
Latinized from Greek *kokkux*, cuckoo, shaped like a cuckoo's bill, as in *Coccyzus minor*, the Mangrove Cuckoo

Coccothraustes kock-ko-THRAW-steez
Cocco, seed, and *thraustes*, to eat, as in *Coccothraustes coccothraustes*, the Hawfinch

Cochlearius koke-lee-AR-ee-us
Cochlear, spoon or spoonful, as in *Cochlearius cochlearius*, the Boat-billed Heron, with a large spoon-shaped bill

Coerulescens seh-roo-LES-senz
Bluish, becoming bluish, as in *Aphelocoma coerulescens*, the Florida Scrub Jay

Colaptes ko-LAP-teez
Latinized from Greek, *kolapto*, to chisel or peck, as in *Colaptes auratus*, the Northern Flicker

Colchicus kol-KEE-kus
After the ancient country of Colchis on the Black Sea where *Phasianus colchicus*, the Common Pheasant, originated

Colaptes auratus, Northern Flicker

Cinnyris

There are 132 species of sunbird; of the 15 genera, *Cinnyris* (SIN-ni-ris) is the largest, with about 45 species. They are typically very small and colorful birds, found in Africa, southern Asia, parts of the Middle East, and the northern tip of Australia. Their main food is nectar, but they supplement their diet with insects for protein when raising young and occasionally eat fruit. They are Old World ecological equivalents of the New World hummingbirds, one main difference being that the sunbirds are passerines (songbirds of the order Passeriformes) while the hummingbirds are in their own order, Apodiformes, along with swifts. *Cinnyris* got its name from the Greek Hesychius of Alexandria who called some unknown bird *kinnuris*.

Unlike hummingbirds, which typically hover when feeding on nectar, sunbirds usually feed from a perch. They have long, curved bills to reach down into the corolla of flowers, but when the corolla tube is too long, they use their bill to puncture the base of the flower. The tongue is extra long, able to project far past the tip of the bill and rolls up from the edges, forming a kind of straw. The end of the tongue is split and jagged on the edges and serves to sop up the nectar, which is drawn up the tubular tongue by capillary action. The very colorful males have longer bills and tongues than the much plainer females, presumably enabling the sexes to exploit different flower sources for nectar.

Cinnyris ludovicensis, Ludwig's Double-collared Sunbird

Sunbirds and hummingbirds represent an example of convergent evolution.

All sunbirds are strikingly beautiful, but only one, the Beautiful Sunbird, gets the name *C. pulchellus*, after the Latin for beautiful, but the Superb Sunbird *C. superbus*, Latin for splendid, superb, and the Regal Sunbird, *C. regius*, Latin for kingly, get the point across. The Eastern Double-collared Sunbird, however, attractive in its own right, does not seem to deserve the unflattering name *C. mediocris*, Latin for ordinary.

Like the similar small-bodied hummingbirds of cold environs, sunbirds that live at high altitudes enter a state of torpor at night to preserve their stored energy. The Southern Double-collared Sunbird, *C. chalybeus*, can lower its body temperature by up to 62.6°F (17°C).

Cinnyris coquerellii, Mayotte Sunbird

COLUMBA

Aristophanes gave the name *kolumbis*, diver, to the Rock Dove or Pigeon, *C. livia*. The Latinized form of the Greek genus *Columba* (ko-LUM-ba) means dove or pigeon. It may refer to their flight behavior, swooping, and diving as if swimming. The terms "dove" and "pigeon" do not denote any real biological difference, although those named pigeons tend to be larger birds. Old English dufe, dive, gives us dove, and pigeon derives from Old French pigeon, meaning young dove.

The *Columba* genus contains 35 species out of 305 species in the family Columbidae, found worldwide except for the extreme south and north, and the driest areas of the Sahara. *Columba* species are mainly Old World, but *C. livia*, the Rock Dove, has been introduced virtually everywhere. The specific epithet *livia* comes from the Latin *livor*, bluish, referring to the bird's grayish-blue coloration.

The Rock Dove has had an amazing history as a messenger from ancient Greece until the end of World War II. Pigeons were an important form of communication during World War I as telegraph lines were not complete. One pigeon, called Cher Ami, delivered a message that saved allied troops and was awarded the Croix de Guerre. For a long time doves have appeared in Olympic opening ceremonies and they have been domesticated and bred into about 200 forms, including racers, homing pigeons, rollers, tumblers, highflyers, and pouters.

Members of *Columba*, like most doves/pigeons, eat seeds, fruits, plant parts, and invertebrates. While almost all other birds have to tilt their heads up to drink by letting gravity put water down their throats, pigeons can suck up water to drink. They typically lay only two eggs and, when the young hatch, feed them with pigeon milk, the sloughed-off lining of a part of the esophagus called the crop. Like many birds, doves have no gall bladder; because they produced no bile, early naturalists surmised that the birds must have a sweet disposition.

Unlike mammals birds have no sweat glands, so they depend on their circulatory and respiratory systems to dissipate excess body heat. They pant or vibrate their throat (called gular fluttering) and lose some heat through the skin. Doves also have a unique plexus of veins and arteries around their esophagus; when the bird is stressed, it expands and deflates its esophagus, transferring heat from the plexus to the esophagus where it escapes through evaporative cooling.

Columba livia, Rock Dove or Pigeon

Columba delegorguei, Eastern Bronze-naped Pigeon

Thirty percent of *Columba* species are considered threatened or near-threatened. Habitat loss is a big factor but some pigeons and doves are hunted for food or sport.

Colinus *ko-LEE-nus*
Latinized form of *zolin*, a Native American word for partridge, as in *Colinus cristatus*, the Crested Bobwhite

Collaris *kol-LAR-is*
Collar, as in *Aythya collaris*, the Ring-necked Duck

Columba *ko-LUM-ba*
Pigeon or dove, perhaps derived from its call, as in *Columba livia*, the Rock Dove or Pigeon. Pigeon derives from the French and dove from the Anglo-Saxon, but there is no biological difference between the two

Columbigallina *ko-lum-bi-gal-LIN-na*
Columbi-, pigeon and *gallina*, hen or cock, as in *Columbigallina minuta*, the Plain-breasted Ground Dove, perhaps because of the way the bird walks on the ground

Columbina *ko-lum-bi-na*
Dove-like as in *Columbina passerina*, the Common Ground Dove

Concolor *KON-ko-lor*
Of one color, as in *Corythaixoides concolor*, the all-gray Gray Go-Away Bird

Concreta *kon-KREE-ta*
Actual, large, strong, as in *Platysteira concreta*, the Yellow-bellied Wattle-eye

Contopus *kon-TOE-pus*
Greek, *kontos*, short, and *pous*, foot as in *Contopus lugubris*, the Dark Pewee

Conuropsis *kon-ur-OP-sis*
Conurus, a genus of Old World parakeets, and *opis*, appearing like, as in *Conuropsis carolinensis*, the Carolina Parakeet, but it was an error in taxonomy

Cooperi *KOO-per-eye*
After William C. Cooper, one of the founders of the American Museum of Natural History in New York, as in *Accipiter cooperii*, Cooper's Hawk

Copsychus *kop-SIK-us*
Greek, *kótsyfas*, blackbird or thrush, as in *Copsychus albospecularis*, the Madagascan Magpie-Robin

Coracina *kor-a-SEEN-a*
Corax, raven, and *-ina*, small, as in *Coracina melanoptera*, the Black-headed Cuckooshrike

Corax *KO-raks*
Raven, as in *Corvus corax*, the Northern Raven

Corniculata *kor-ni-koo-LA-ta*
Corn, horn, and *culata*, little, piece, as in *Fratercula corniculata*, the Horned Puffin, which has a fleshy black "horn" above each eye

Cornuta *kor-NOO-ta*
Horned, as in *Anhima cornuta*, the Horned Screamer

Coronata *kor-o-NA-ta*
Crowned, as in *Setophaga coronata*, the Myrtle Warbler, with a yellow crown

Coruscans *KOR-us-kanz*
Coruscus, glittering, shining, as in *Neodrepanis coruscans*, the Common Sunbird-Asity

Corvus *KOR-vus*
Crow, as in *Corvus albus*, Pied Crow

Coturnix *ko-TUR-niks*
Quail, as in *Coturnix coturnix*, the Common Quail; the name probably derived from the bird's three short-syllable call

Cracticus *KRAK-ti-kus*
Greek, *kraktikos*, to shriek like a raven, as in *Cracticus mentalis*, the Black-backed Butcherbird

Conuropsis carolinensis, Carolina Parakeet

Corvus

The genus of about 40 species of birds commonly known as crows or ravens, *Corvus (Kor-vus)*, is Latin for "crow." Found almost all over the world except the polar continents and South America, the members of this genus are very adaptable and successful and perhaps the most intelligent of all birds. In the US, the American Crow, *C. brachyrhynchos*, "crow with a short bill", is the most recognizable. In Europe it is either the Carrion Crow (*C. corone*, Latin *corvus*, crow, and Greek, *corone*, raven, hence the crow-raven) or Hooded Crow (*C. cornix*, both Latin terms meaning crow). Other birds in this genus have more descriptive names such as *C. albicollis*, the White-necked Raven.

Crows, ravens, and their relatives in the family Corvidae have gained the deserved reputation of being the smartest of the bird world. They make tools, play games, speak human words, find hidden objects, drop walnuts into road traffic so that cars expose the nuts' innards, bait fish with bread crumbs, and even recognize individual human faces. The New Caledonian Crow, *C. moneduloides*, the most intelligent of any bird, uses tools and can make a hook (for hooking insects, fruits, or nuts from crevices), something even our nearest relatives, the chimpanzees, cannot do.

Another reason for their success is their diet. They will eat almost anything, animal or vegetable, alive or dead. This foraging habit is called euryphagy (wide diet). They also have a high tolerance for human activity and are occasionally crop pests, a habit that stimulated the invention of the scarecrow some 3,000 years ago.

Ravens and crows appear frequently in mythology and folklore. There are several Native American stories about how the crow (or raven) became black after rescuing the moon, sun, and stars from an owl's lair. In Norse mythology, two ravens roamed the world to bring information back to the king. And there is a British superstition that if ravens ever fled the Tower of London, the monarchy would fall, so six ravens are kept there and overseen by Yeoman Warders.

Partly because of their black color, crows and ravens have often been considered bad omens, foretelling or announcing deaths and perhaps giving rise to the name "murder" for a group of crows.

Corvus brachyrhynchos, American Crow

Corvus corax, Northern Raven

Ravens occasionally play by repeatedly rolling down a snow-covered hill or house roof.

Crassirostris *kras-si-ROSS-tris*
Thick-billed, as in *Corvus crassirostris*, the Thick-billed Raven

Creatopus *kree-a-TOE-pus*
Greek, *creas*, flesh, and *pous*, foot, as in *Puffinis creatopus*, the Pink-footed Shearwater

Crecca *KREK-ka*
A Latinized word meant to express the sound of the bird, as in *Anas crecca*, the Eurasian Teal

Crinitus *KRIN-ih-tus*
Crinit, hair or bearded, probably referring to the moveable crest, as in *Myiarchus crinitus*, the Great Crested Flycatcher

Cristata *kris-TA-ta*
Crested, as in *Gubernatrix cristata*, the Yellow Cardinal, and *Cyanocitta cristata*, the Blue Jay

Cristatus *kris-TA-tus*
Crested, as in *Colinus cristatus*, the Crested Bobwhite and the Peacock, *Pavo cristatus*

Crocethia *krow-SETH-ee-a*
Greek, stone or pebble chaser or runner, as in *Crocethia* (now *Calidris*) *alba*, the Sanderling, known for its habit of running back and forth at the edge of the surf

Crotophaga *kro-tow-FAY-ga*
Greek, *kroton*, tick, insect, and *phago*, eat, as in *Crotophaga ani*, the Smooth-billed Ani, which feeds not only on insects but on seeds and fruit

Cuculus *koo-KOO-lus*
A cuckoo, from the Common Cuckoo's call, as in *Cuculus canorus*

Cunicularia *koo-ni-koo-LAR-ee-a*
Cunicul, an underground passage, as in *Athene cunicularia*, the Burrowing Owl, which nests in an underground burrow, either dug itself, soil permitting, or made by a mammal

Cuvieri, -ii *koo-vee-AIR-eye/ee-eye*
After Georges Cuvier, a French naturalist, as in *Dryolimnas cuvieri*, the White-throated Rail

Cyaneoviridis *sye-an-ee-o-vi-RI-dis*
Cyaneus, dark blue, and *viridis*, green, as in *Tachycineta cyaneoviridis*, the Bahama Swallow

Cyanocephalus, -a *sye-an-o-se-FAL-us/a*
Cyaneus, dark blue, and *cephala*, head, as in *Euphagus cyanocephalus*, Brewer's Blackbird

Cyanocitta *sye-an-o-SIT-ta*
Cyaneus, dark blue, and Greek *kitta*, jay, as in *Cyanocitta cristata*, the Blue Jay

Cyanocorax *sye-an-o-KOR-aks*
Cyaneus, dark blue and Greek, *corax*, raven, as in *Cyanocorax chrysops*, the Plush-crested Jay

Cyanogaster *sye-an-o-GAS-ter*
Cyaneus, dark blue, and Greek, *gaster*, stomach, as in *Coracias cyanogaster*, the Blue-bellied Roller

Cyanomelana *sye-an-o-mel-AN-a*
Cyaneus, dark blue and Greek, *melas*, black, as in *Cyanoptila cyanomelana*, the Blue-and-white Flycatcher

Cyanoptera *sye-an-OP-ter-a*
Cyaneus, dark blue, and Greek, *pteron*, wing, as in *Anas cyanoptera*, the Cinnamon Teal with its blue wing patches

Cyanoptila *sigh-an-op-TIL-a*
Cyaneus, dark blue, and Greek, *pteron*, wing, as in *Cyanoptila cumatilis*, Zappey's Flycatcher

Cygnus *SIG-nus*
Greek, *kuknos*, swan, as in *Cygnus atratus*, the Black Swan

Cyrtonyx *sir-TON-iks*
Greek, *kurtos*, bent, and Latin *onyx*, claw, as in *Cyrtonyx ocellatus*, the Ocellated Quail, for the sickle-shaped claws it uses for digging

Tachycineta cyaneoviridis, Bahama Swallow

D

Dactylatra dak-til-AH-tra
Greek, *dactyl*, finger or toe, and Latin, *ater*, dark or black, as in *Sula dactylatra*, the Masked Booby, from black primary feathers. Booby comes from the Spanish *bobo*, referring to a slow-witted person or ungainly bird

Dactylortyx dak-til-OR-tiks
Greek, *dactyl*, finger or toe, and *ortux*, quail, as in *Dactylortyx thoracicus*, the Singing Quail

Damophila dam-o-FIL-a
Greek, poetess, contemporary with Sappho, as in *Damophila julie*, the Violet-bellied Hummingbird

Daption DAP-tee-on
An anagram of Pintado, as in *Daption capense*, the Cape Petrel, also called the Pintado Petrel

Daptrius DAP-tree-us
Greek, *daptes*, eater, to devour, as in *Daptrius ater*, the Black Caracara, a predatory bird

Darwini, -ii DAR-win-eye/dar-WIN-ee-eye
After Charles Darwin, British naturalist and explorer, who famously observed the Galapagos Finches, now called Darwin's Finches, although none of them has a genus or species name *darwini* as does *Nothura darwinii*, Darwin's Nothura, a type of tinamou

Dasyornis das-ee-OR-nis
Greek, *dasus*, hairy, shaggy, and *ornis*, bird, as in *Dasyornis brachypterus* the Eastern Bristlebird

Davidi DA-vi-dye
After Andre David-Beaulieu, a naturalist in French Indo-China, as in *Arborophila davidi*, Orange-necked partridge; also after Pierre David, French priest and zoologist, as in *Strix davidi*, Pére David's Owl

Davisoni DAY-vi-son-eye
After William Davison, curator of Raffles Museum in Singapore, as in *Pseudibis davisoni*, White-shouldered Ibis

Deconychura de-con-ih-KOO-ra
Greek, *deca-*, ten, *onux*, claw, and *oura*, tail, as in *Deconychura longicauda*, the Long-tailed Woodcreeper; refers to the bird's ten tail feathers as "claws" that help the bird climb

Paradisaea decora, Goldie's Bird of Paradise

Decora dek-OR-a
Elegant, as in *Paradisaea decora*, Goldie's Bird of Paradise, after Andrew Goldie, who discovered the bird in 1882

Deglandi DEG-land-eye
After Côme Degland, French ornithologist, who published *European Ornithology* in 1849, as in *Melanitta deglandi*, the White-winged Scoter

Delawarensis del-a-ware-EN-siss
After the Delaware River on the Atlantic Coast of the US, as in *Larus delawarensis*, the Ring-billed Gull that was first described there

Deleornis del-ee-OR-nis
Greek, *dele-*, visible, and *ornis*, bird, as in *Deleornis fraseri*, Fraser's Sunbird

Delicata del-ih-KA-ta
Pleasing, alluring, as in *Gallinago delicata*, Wilson's Snipe, after Scottish-American ornithologist Alexander Wilson

Delothraupis del-o-THRAW-pis
Greek, *delas*, visible, and *thraupis*, tanager, as in *Delothraupis castaneoventris*, the Chestnut-bellied Mountain Tanager

Deltarhynchus del-ta-RINK-us
Greek, *delta*, the letter D, and Latin, *rhynchus*, bill, as in *Deltarhynchus flammulatus*, the Flammulated Flycatcher. The D comes from the triangular shape of the bill in cross-section as in the Greek letter Δ

Demigretta dem-ee-GRET-ta
Old French, *demi*, half or half-sized. *Demigretta* was changed to *Egretta*, as in *Egretta sacra*, the Pacific Reef Heron, which is much smaller than the Great Egret; egret comes from the old French *aigrette*, referring to feather tufts or plumes

Dendragapus den-dra-GAP-us
Greek, *dendron*, tree and *agapo*, to love, be fond of, as in *Dendragapus obscurus*, the Dusky Grouse

Dendrexetastes den-dreks-eh-TAS-teez
Greek, *dendron*, tree and *exetastes*, inspectors or examiners, as in *Dendrexetastes rufigula*, the Cinnamon-throated Woodcreeper

Dendrocincla den-dro-SINK-la
Greek, *dendron*, tree, and *cincla*, circling, as in *Dendrocincla turdina*, the Plain-winged Woodcreeper, so named because of its habit of circling trees while moving up the trunk

Dendrocitta den-dro-SIT-ta
Greek, *dendron*, tree and *citta*, a jay or chattering bird, as in *Dendrocitta frontalis*, the Collared Treepie, pie from the Latin *pica*, meaning magpie

Dendrocolaptes den-dro-ko-LAP-teez
Greek, *dendron*, tree and *colapte*, to chisel, peck out, as in *Dendrocolaptes picumnus*, the Black-banded Woodcreeper

Dendrocopos den-dro-KOPE-os
Greek, *dendron*, tree and *kopis*, to split or cleave, as in *Dendrocopos major*, the Great Spotted Woodpecker

Dendrocygna den-dro-SIG-na
Greek, *dendron*, tree and *cygn*, swan, as in *Dendrocygna bicolor*, the Fulvous Whistling Duck, which sometimes nests in trees

Dendroica den-DROY-ka
Greek, *dendron*, tree, and *oikos*, home or habitat, as in *Dendroica* (now *Setophaga*) *cerulea*, the Cerulean Warbler

Dendronanthus den-dro-NAN-thus
Greek, *dendron*, tree, and Latin, *anthus*, a flower or a lark, as in *Dendronanthus indicus*, the Forest Wagtail (see box), which resembles a lark

Dendropicos den-DRO-pi-kos
Greek, *dendron*, tree, and Spanish, *pico*, small, sharp, beak, as in *Dendropicos lugubris*, the Melancholy Woodpecker

Dendrortyx den-DROR-tiks
Greek, *dendron*, tree, and *ortux*, quail, as in *Dendrortyx barbatus*, the Bearded Wood Partridge

Denhami DEN-am-eye
After Dixon Denham, English soldier and explorer, as in *Neotis denhami*, Denham's Bustard

Diadema dye-a-DEM-a
Greek, crown or turban, as in *Charmosyna diadema*, the probably extinct New Caledonian Lorikeet with a deep blue crown

Diademata, -us dee-a-dem-AH-ta/tus
Diadema, crowned, as in *Alethe diademata*, the White-tailed Alethe, which has a crest that it can raise

Diardi dee-AR-dye
After Pierre Diard, a French explorer and collector in the East Indies, as in *Lophura diardi*, Siamese Fireback

LATIN IN ACTION

The Forest Wagtail of eastern Asia is part of the wagtail, pipit, and longclaw family, Motacillidae, but in its own genus because of some unique characteristics. Most wagtails move their tails up and down; the Forest Wagtail wags sideways. Most members of Motacillidae inhabit open areas, eat insects, and nest on the ground. The Forest Wagtail nests in trees in the forest. In Sri Lanka, the birds feed on maggots in cattle dung.

Dendronanthus indicus, Forest Wagtail

Bird Beaks

Beaks are a defining characteristic of birds. Since most birds use their feet predominantly for walking or perching, and wings for flying or swimming, the primary tool they use to build nests, to find, capture, and eat food, and to defend themselves is their beak, or bill. Beaks are used to court, to make sounds, and to filter water for food. Their shape is thus a reflection of the birds' lifestyle and an excellent characteristic for identification in the field. The Greek suffixes *-rhino*, *-rostrum*, and *-rhyncho* are often used in scientific names to indicate a bill's shape or color, as in the Rhinoceros Hornbill—*Buceros* (horned) *rhinoceros* (nose horn)—and Mallard—*Anas* (duck) *platyrhynchos* (flat bill).

The bill is covered by a fibrous structural protein layer called the rhamphotheca (literally, bill case), the same protein that makes up the outer layer of human skin as well as hair and nails. The rhamphotheca grows continually to counter the wear on the bill. The tips and edges of the bill are supplied with nerve endings to allow the bird to feel what it is touching and to manipulate it. In long-billed shorebirds the tip of the bill is well supplied with sensory cells so that they can find food among sand and gravel, and the tip of the bill can open without the entire jaw opening against the substrate.

The shape of a bill is largely determined by the food needs of the bird. Flycatchers that snatch their prey in mid-air have a flattened triangle-shaped bill with a hook at the end to hold larger prey items, as does the Blue-billed Black Tyrant, *Knipolegus cyanirostris*. Nighthawks, swifts, and swallows have small bills but large mouths with sticky membranes to capture insects. Sparrows have heavy bills with palates especially designed to crack open seeds. Sunbirds and hummingbirds have long, thin, tubular bills to fit into the corolla of flowers, and birds like shoveler ducks (*Anas clypeata*, from *clypeatus*, shield, referring to the shape of the bill) have bills with lamellae that filter out food items from water or mud. The American Avocet, *Recurvirostra americana*, has a recurved bill to help it skim invertebrates from the

Amazona guildingii, *Aulacorhynchus prasinus*, *Rhinoplax vigil*, St. Vincent Amazon, Emerald Toucanet, Helmeted Hornbill

Although not the most attractive feature of some birds, the bill can be important in attracting a mate during the breeding season.

BIRD BEAKS

Rynchops niger,
Black Skimmer

At hatching, the young Black Skimmer has equal-sized upper and lower bills, but by the time they leave the nest (fledge) the lower bill is a bit longer.

Since beaks are the major anatomical feature determining a bird's niche, they often reduce competition, either within or between species, by being slightly different. The male bill may differ from that of the female just enough in either size or shape to enable them to eat somewhat different food items. The classic example is Darwin's finches of the Galapagos Islands. There are 13 species but only a few on each island. The set of species on each island evolved beaks that were somewhat different in size or shape than the other species to exploit the food resources on their island more effectively. And any one species will look somewhat different than its conspecific cohorts on other islands. The one with the largest bill, is, of course, *Geospiza magnirostris*, the Large Ground Finch.

So, by looking at a bird's bill, you can often deduce a lot about its lifestyle.

surface of the water. Toucan bills are large and long, supposedly for reaching fruits in thick vegetation, but recent research has demonstrated that they are also a thermoregulatory mechanism, using a blood supply to increase or diminish the bird's body temperature. Bills are usually distinctive, as in the Spot-billed Toucanet, *Selenidera maculirostris*, the species name referring to the spotted bill.

The Black (*Rynchops niger*), Indian (*R. albicollis*) and African Skimmers (*R. flavirostris*) have a unique bill with the lower jaw being longer than the upper one. They skim along ocean shores with their lower bill in the water and when they feel a fish, crustacean, or mollusk such as a squid, they snap their bills shut. The lower bill's ramphotheca, subject to this wear and tear, grows faster than the upper bill's. Skimmers also have the only slit-shaped pupils in the bird world, enabling them to see their bill tip.

Buceros rhinoceros,
Rhinoceros Hornbill

The male walls the female in a tree cavity by layering mud over the entrance, leaving only a small hole through which he passes food to her while she incubates the eggs.

Diazi dee-AZ-eye
After Augustin Diaz, Mexican engineer and director of the Mexican Geographical and Exploring Commission, as in *Anas diazi*, the Mexican Duck

Dichroa dye-KRO-a
Greek, *di-*, two or separate, and *chroa*, color, as in *Aplonis dichroa*, the Makira Starling, after its blue-black iridescent coloration

Dichromanassa dye-kro-ma-NASS-sa
Greek, *di-*, two, *chrom*, color, and *anassa*, queen, as in *Dichromanassa* (now *Egretta*) *rufescens*, the Reddish Egret, named for its two color phases, reddish and white

Dichrous DYE-krus
Greek, *di-*, two or separate, and *chrous*, color or complexion, as in *Pitohui dichrous*, the Hooded Pitohui, a bird that, along with a couple of close relatives, accumulates poison in its skin from eating a certain beetle

Dichrozona dye-kro-ZONE-a
Greek, *di-*, two, *chrous*, color or complexion, *zona*, a belt, girdle, zone, as in *Dichrozona cincta*, the Banded Antbird

Dicrurus dy-KROO-rus
Greek, *dicros*, forked and *ourus*, tail, as in *Dicrurus forficatus*, the Crested Drongo, drongo from the local name given to it by Madagascar natives

Discosura conversii, Green Thorntail

Difficilis dif-fi-SIL-is
Difficult, as in *Empidonax difficilis*, the Pacific Slope Flycatcher, and probably referring to the fact that *Empidonax* species are difficult to tell apart

Diglossa dye-GLOS-sa
Greek, *di-*, two, and *glossa*, tongue, as in *Diglossa humeralis*, the Black Flowerpiercer, referring to its fringed tongue

Diglossopis dye-glos-SO-pis
Greek, two-tongued, as in *Diglossopis cyanea*, the Masked Flowerpiercer. *Diglossopis* is often subsumed under *Diglossa*

Dinopium di-NO-pee-um
Greek, *dinos*, terrible, whirling, and *ops*, appearance, as in *Dinopium benghalense*, the Black-rumped Flameback, the genus name apparently referring to its unusually large size for a woodpecker

Diomedea dye-o-meh-DEE-a
After Diomedes, hero of the Trojan War whose companions were turned into birds, as in *Diomedea exulans*, the Wandering Albatross

Diophthalma dy-op-THAL-ma
Greek, *di-*, two and *opthalmos*, eye, as in *Cyclopsitta diophthalma*, the Double-eyed Fig Parrot; the cheek patches of some subspecies resemble eyespots

Diops DYE-ops
Greek, *di-*, two and *ops*, appearance, face, or eyes as in *Todiramphus diops*, the Blue-and-White Kingfisher

Diopsittaca dye-op-SIT-ta-ka
Greek, *dio*, divine, noble, and *psittaca*, parrot, as in *Diopsittaca nobilis*, the Red-shouldered Macaw

Diphone dye-FO-nee
Greek, *di-*, two, and *phone*, sound or voice, as in *Horornis diphone*, the Japanese Bush-warbler, whose beautiful voice is heard far more often than the bird is seen

Discolor DIS-ko-lor
Greek, *dis-*, separate, Latin, *color*, color, as in the different colors of *Certhia discolor*, the Sikkim Treecreeper; one population of this bird in Burma/Myanmar has a brown throat and has been considered a subspecies, though some consider it a separate species

Discors DIS-korz
Discordant, disagreeable, as in *Anas discors*, the Blue-winged Teal; *discors* might refer to its call, the noise it makes while taking off, or its face or wing patterns

Discosura dis-ko-SOO-ra
Greek, *dis-*, apart, separate, and *oura*, tail, as in *Discosura conversii*, the Green Thorntail, whose highly modified tail feathers give it its name

Discurus dis-KOO-rus
Greek, *disc*, a round plate, and *oura*, tail, as in *Prioniturus discurus*, the Blue-crowned Racket-tail

Disjuncta dis-JUNK-ta
Separate, disjunct, as in *Myrmeciza disjuncta*, the Yapacana Antbird, Yapacana an area of Venezuela. *Disjuncta* refers to the unresolved taxonomy of this bird in relation to other *Myrmeciza* species

Dissimilis dis-SIM-ih-lis
Dissimilar, as in *Turdus dissimilis*, the Black-breasted Thrush; most thrushes exhibit little or no sexual dimorphism, but *Turdus dissimilis* does

Dixiphia diks-ih-FEE-a
Greek, *di-*, two, and *xiphos*, sword, as in *Dixiphia pipra*, the White-crowned Manakin. The anatomy of its syrinx (voice box) resembles two crossed swords

Dohertyi doe-ERT-ee-eye
After William Doherty, an American collector of insects and later birds, as in *Ptilinopus dohertyi*, Red-naped Fruit Dove

Dohrnii DORN-ee-eye
After Felix Dohrn, German founder of the first zoological research station in the world, the Stazione Zoologica in Naples, Italy, as in *Glaucis dohrnii*, the Hook-billed Hermit, a type of hummingbird

Dolei DOL-eye
After Sanford Dole, a lawyer and jurist in the Hawaiian Islands, as in *Palmeria dolei*, the Akohekohe

Dolichonyx doe-li-KON-iks
Greek, *dolichos*, long, and *onux*, claw as in *Dolichonyx oryzivorus*, the Bobolink, referring to its long claws. Bobolink derives from bob-o-lincoln, imitative of its call

Doliornis doe-lee-OR-nis
Greek, *dolio*, shrewd, and *ornis*, bird, as in *Doliornis remseni*, the Chestnut-bellied Cotinga. Probably from its secretive habits; it was not discovered until 1989

Domesticus doe-MESS-ti-kus
Domesticus, around the house, as in *Passer domesticus*, the House Sparrow

Ptilinopus dohertyi,
Red-naped Fruit Dove

Dominica, -cana, -canus, -censis doe-MIN-ih-ka/doe-min-ih-KAN-a/kan-us, doe-min-ih-SEN-sis
After the Commonwealth of Dominica in the West Indies, as in *Pluvialis dominica*, the American Golden Plover, which passes through the West Indies during migrations

Donacobius don-a-KO-bee-us
Greek, *donax*, a reed, and *bios*, life, living, as in *Donacobius atricapilla*, the Black-capped Donacobius, which inhabits reeds and other vegetation in wetlands of the Amazonian basin

Donacospiza don-a-ko-SPY-za
Greek, *donax*, a reed, and *spiza*, finch, as in *Donacospiza albifrons*, the Long-tailed Reed Finch

Donaldsoni DON-ald-son-eye
After Arthur Donaldson-Smith, an American traveler, African big-game hunter, and Fellow of the Royal Geographical Society, as in *Caprimulgus donaldsoni*, Donaldson-Smith's Nightjar

Dorsalis, -ae dor-SAL-is/ee
Dorsum, back, from the back, as in *Mimus dorsalis*, the Brown-backed Mockingbird

Dorsimaculatus dor-si-mak-oo-LAT-us
Dorsum, back, and *macula*, spot, as in *Herpsilochmus dorsimaculatus*, the Spot-backed Antwren

Dorsomaculatus dor-so-mak-oo-LA-tus
Dorsum, back, and *macula*, spot, as in *Ploceus dorsomaculatus*, the Yellow-capped Weaver (with a spotted back)

LATIN IN ACTION

The antbirds, such as *Drymophila malura*, the Dusky-tailed Antbird, found in Central and South America, belong to a couple of different families with over 200 species. They do not eat ants but hunt a variety of other arthropods such as mantids, cockroaches, beetles, bees, and so forth, by hopping through the brush or snatching prey in flight. Some species follow Army Ants and as the ants dislodge arthropods or similar prey items from their hiding places, the birds pounce upon them. As these birds resemble other bird families, we find names such as antthrushes, antvireos, antshrikes, and antpittas. The birds will also crush ants and rub them through their feathers as ants' formic acid deters feather parasites.

Dorsostriatus *dor-so-stree-AT-us*
Dorsums, back, *striatus*, striated, striped, as in *Serinus dorsostriatus*, the White-bellied Canary

Dougallii *DOO-gal-eye*
After Peter McDougall, Scottish physician and naturalist, as in *Sterna dougallii*, the Roseate Tern

Drepanis *dre-PAN-is*
Greek, *drepane*, sickle, as in *Drepanis pacifica*, the now extinct Hawaii Mamo, referring to its downcurved bill

Drepanoptila *dre-pan-OP-til-a*
Greek, *drepane*, sickle, and *ptil-*, feather, as in *Drepanoptila holosericea*, the Cloven-feathered Dove

Drepanorhynchus *dre-pan-o-RINK-us*
Greek, *drepane*, sickle, and Latin, *rhynchus*, bill, as in *Drepanorhynchus reichenowi*, the Golden-winged Sunbird

Dromas *DRO-mas*
Greek, *dromas*, run, running a race, as in *Dromas ardeola*, the Crab Plover

Dromococcyx *dro-mo-KOK-siks*
Greek, *dromas*, run, running a race, and *coccyx*, cuckoo, as in *Dromococcyx pavoninus*, the Pavonine Cuckoo; *Pavoninus* is Latin for resembling a peacock

Dryas *DRY-as*
Dryad, tree or wood-nymph, as in *Catharus dryas*, the Spotted Nightingale-Thrush

Drymocichla *dry-mo-SICK-la*
Greek, *drymo*, woodland, forest, and *cichla*, thrush, as in *Drymocichla incana*, the Red-winged Gray Warbler

Drymodes *dry-MO-deez*
Greek, *drymo*, woodland, forest, as in *Drymodes brunneopygia*, the Southern Scrub Robin

Drymophila *dry-mo-FIL-a*
Greek, *drymo*, woodland, forest, and *philos*, like, love, as in *Drymophila malura*, the Dusky-tailed Antbird

Drymornis *dry-MOR-nis*
Greek, *drymo*, woodland, forest, and *ornis*, bird, as in *Drymornis bridgesii*, the Scimitar-billed Woodcreeper

Dryolimnas *dry-o-LIM-nas*
Greek, *drus*, tree, and Latin, *limnas*, marsh or lake, as in *Dryolimnas cuvieri*, the White-throated Rail

Dryoscopus *dry-o-SKO-pus*
Greek, *drus*, tree, and *skopus*, look, watcher, as in *Dryoscopus cubla*, the Black-backed Puffback, with fluffy upper tail coverts

Dubia *DOO-bee-a*
Doubtful, uncertain, as in *Alcippe dubia*, the Rusty-capped Fulvetta, whose taxonomic relationship is uncertain

Dubius *DOO-bee-us*
Doubtful, uncertain, as in *Lybius dubius*, the Bearded Barbet, so named because of early confusion as to the taxonomy of the bird

Ducula *doo-KOO-la*
To lead, as in *Ducula bakeri*, the Vanuatu Imperial Pigeon

Duidae *doo-EE-dee*
After Cerro Duida, a mountain in Venezuela, as in *Diglossa duidae*, the Scaled Flowerpiercer

Dumetella *doo-meh-TEL-la*
Dumetum, shrub, bramble, and *ella*, diminutive, as in *Dumetella carolinensis*, the Gray Catbird, which inhabits brushy areas

Dumetia *dum-ET-ee-a*
Shrub, bramble, as in *Dumetia hyperythra*, the Tawny-bellied Babbler

E

Eatoni *EE-ton-eye*
After Alfred Eaton, English explorer and naturalist, as in *Anas eatoni*, Eaton's Pintail

Eburnea *ee-BUR-nee-a*
Eburne, ivory, as in *Pagophila eburnea*, the Ivory Gull

Ecaudatus *eh-kaw-DA-tus*
E-, without, and *caudata*, tail, as in *Myiornis ecaudatus*, the Short-tailed Pygmy Tyrant, with a stub of a tail; it is also the smallest passerine (songbird) in the world

Ectopistes *ek-toe-PIS-teez*
Greek, *ectopistes*, wanderer, as in *Ectopistes migratorius*, the extinct Passenger Pigeon, the common name coming from French settlers who called the bird "*Pigeón de passage*," pigeon of passage

Edwardsi *ED-wards-eye*
Probably after George Edwards, an English naturalist and ornithologist, the "father of British ornithology," as in *Bangsia edwardsi*, the Moss-backed Tanager

Edwardsii *ed-WARDS-ee-eye*
After Alphonse Milne-Edwards, a renowned French naturalist, as in *Carpodacus edwardsii*, the Dark-rumped Rosefinch

Egertoni *EJ-er-ton-eye*
After Sir Philip Egerton, an English paleontologist and member of the House of Commons, as in *Actinodura egertoni*, the Rusty-fronted Barwing

Egregia *ee-GREE-gee-a*
Egregia, distinguished, as in *Crex egregia*, the African Crake. The species name may refer to its upright distinguished posture and crake after its call

Egretta *ee-GRET-ta*
Old French, *aigrette*, a kind of heron, as in *Egretta vinaceigula*, the Slaty Egret. There is no biological difference between herons and egrets

Eichhorni *IKE-horn-eye*
After Alfred Eichhorn, an Australian farmer, as in *Philemon eichhorni*, the New Ireland Friarbird

Eisentrauti *EY-zen-trout-eye*
After Martin Eisentraut, German zoologist and collector, as in *Melignomon eisentrauti*, the Yellow-footed Honeyguide

Elachus *ee-LAK-us*
Greek, *elach*, small, as in *Dendropicos elachus*, the Little Gray Woodpecker

Elaenia *eh-LEEN-ee-a*
Greek, *elaeo*, olive, olive oil, as in *Elaenia parvirostris*, the Small-billed Elaenia, a tyrant flycatcher

Elanoides *el-a-NOY-deez*
Elanus, kite, and Greek, *eidos*, like, resemble, as in *Elanoides forficatus*, the Swallow-tailed Kite

Elanus *eh-LAN-us*
Elanus, kite, as in *Elanus leucurus*, the White-tailed Kite, whose common name derives from the string-held toy

Elaphrus *ee-LAF-rus*
Greek, *elaphros*, light in weight, as in *Aerodramus elaphrus*, the Seychelles Swiftlet

Elata, -us *ee-LAY-ta/tus*
Elat, high, lofty, as in *Ceratogymna elata*, the Yellow-casqued Hornbill

Ectopistes migratorius, Passenger Pigeon

Electron ee-LEK-tron
Greek, *electr-*, amber, electricity, as in *Electron platyrhynchum*, the Broad-billed Motmot; refers to the color of the head and chest, while Motmot derives from the call

Elegans EL-le-ganz
Elegantem, choice, fine, tasteful, as in *Pitta elegans*, the Elegant Pitta, and about 20 other species' names

Elegantissima eh-le-gan-TISS-see-ma
Very elegant, as in *Euphonia elegantissima*, the Elegant or Blue-headed Euphonia

Eleonorae el-lee-o-NOR-ee
After Eleanor of Arborea, national heroine of Sardinia, as in *Falco eleonorae*, Eleonora's Falcon

Ellioti, -ii EL-lee-ot-eye/el-lee-OT-ee-eye
After Daniel Elliot, Curator of Zoology at the Field Museum in Chicago, as in *Syrmaticus ellioti*, Elliot's Pheasant

Elseyornis el-see-OR-nis
After Joseph Elsey, English surgeon, explorer, and naturalist, and Greek, *ornis*, bird, as in *Elseyornis melanops*, the Black-fronted Dotterel, *dotterel* from Middle English, meaning stupid, silly

Emberiza em-be-RYE-za
Swiss-German, *emmeritz*, bunting, as in *Emberiza cia*, the Rock Bunting; there is no clear etymology for bunting

Emberizoides em-ber-ih-ZOY-deez
Swiss-German, *emmeritz*, bunting, and Greek, *oid*, like, resembling, as in *Emberizoides ypiranganus*, the Lesser Grass Finch

Emblema em-BLEM-a
Inlaid ornamental work, as in *Emblema pictum*, the Painted Finch

Pitta elegans, Elegant Pitta

Enodes erythrophris, Fiery-browed Starling

Empidonax em-pi-DON-aks
Greek, *empis*, gnat, mosquito, and *anax*, king, as in *Empidonax flaviventris*, the Yellow-bellied Flycatcher. There are about 15 *Empidonax* species, many of which are difficult to tell apart, often called "empids" by birdwatchers

Empidonomus em-pi-DON-o-mus
Greek, *empis*, gnat, mosquito, and *nomas*, grazing, as in *Empidonomus varius*, the Variegated Flycatcher

Empidornis em-pi-DOR-nis
Greek, *empis*, gnat, mosquito, and *ornis*, bird, as in *Empidornis semipartitus*, the Silverbird

Endomychura en-do-my-KOO-ra
Greek, *endo*, inner, *mycho*, inward, and *oura*, tail, as in *Endomychura* (now *Synthliboramphus*) *hypoleucus*, the Guadalupe Murrelet, alluding to the very short tail

Enganensis en-ga-NEN-sis
After Enggano Island, Indonesia, as in *Otus enganensis*, the Enggano Scops Owl

Enigma eh-NIG-ma
Puzzle, mystery, as in *Todiramphus enigma*, the Talaud Kingfisher, of the Taulaud Islands, Indonesia

Enodes ee-NO-deez
Smooth, as in *Enodes erythrophris*, the Fiery-browed Starling, with very smooth plumage

Ensifera en-si-FER-a
Ensi, sword, and *fer*, to bear, as in *Ensifera ensifera*, the Sword-billed Hummingbird

Ensipennis en-si-PEN-nis
Ensi, sword, and *pennis*, feather, wing, as in *Campylopterus ensipennis*, the White-tailed Sabrewing

Entomodestes en-toe-mo-DES-teez
Greek, *entomo*, insect, and *edest*, eater, as in *Entomodestes coracinus*, the Black Solitaire

Enucleator ee-noo-clee-AH-tor
E-, without, and *nucleator*, nucleus or seed, as in *Pinicola enucleator*, the Pine Grosbeak, which extracts seeds from pine cones

Eolophus ee-o-LO-fus
Greek, *eo*, dawn, early, and Latin, *lophus*, crest, as in *Eolophus roseicapilla*, the Galah; galah is derogatory Australian slang, meaning fool or idiot

Eophona ee-o-FONE-a
Greek, *eo*, dawn, early, and *phon*, sound, voice, as in *Eophona personata*, the Japanese Grosbeak

Eopsaltria ee-op-SAL-tree-a
Greek, *eo*, dawn, early, and *psalter*, female lyre player, as in *Eopsaltria australis*, the Eastern Yellow Robin

Eos EE-os
Greek, *eo*, dawn, early, as in *Eos histrio*, the Red-and-blue Lory; apparently refers to both the bright red plumage and eastern Indonesia distribution (*Eos* in reference to the sun rising in the east)

Epauletta eh-paw-LET-ta
French, *épaulette*, shoulder ornament, as in *Pyrrhoplectes epauletta*, the Golden-naped Finch

Epichlorus eh-pi-KLOR-us
Greek, *epi-*, on, over, and *chloro-*, green, as in *Urolais epichlorus*, the Green Longtail

Epimachus ep-ih-MAK-us
Greek, *epimakos*, ready for battle, as in *Epimachus meyeri*, the Brown Sicklebill

Episcopus eh-PIS-ko-pus
Episcopus, an overseer or bishop, as in *Ciconia episcopus*, the Woolly-necked Stork or Bishop Stork, because it resembles a religious entity with its white collar

Epops EE-pops
Greek, *epops*, hoopoe, as in *Upupa epops*, the Eurasian Hoopoe; the common name comes from its call

Epulata eh-poo-LAT-a
Epul, feast, and *ata*, full, as in *Muscicapa epulata*, the Little Gray Flycatcher

Eques EH-kweez
A horseman, knight, as in *Myzomela eques*, the Ruby-throated Myzomela

Erckelii er-KEL-ee-eye
After Theodor Erckel, a German taxidermist, as in *Pternistis erckelii*, Erckel's Francolin

Eremalauda eh-rem-a-LAW-da
Greek, *eremos*, a lonely place, and Latin, *alauda*, lark, as in *Eremalauda dunni*, Dunn's Lark, often found in remote desert areas

Eremiornis eh-rem-ee-OR-nis
Greek, *eremos*, a lonely place, and *ornis*, bird, as in *Eremiornis* (now *Megalurus*) *carteri*, the Spinifexbird, *Spinifex* being a genus of grasses in which the bird nests

Upupa epops,
Eurasian Hoopoe

Eremita *eh-ri-MIT-a*
Eremita, hermit, as in *Geronticus eremita*, the Northern Bald Ibis or Hermit Ibis

Eremomela *eh-rem-o-MEL-a*
Greek, *eremos*, a lonely place, and *melo*, song, as in *Eremomela pusilla*, the Senegal Eremomela

Eremophila *eh-re-mo-FIL-a*
Greek, *eremos*, a lonely place, and *philia*, love, as in *Eremophila alpestris*, the Horned or Shore Lark

Ereunetes *eh-re-un-EET-eez*
Greek, *ereunetes*, prober, as in *Ereunetes* (now *Calidris*) *pusilla*, the Semipalmated Sandpiper, which probes for invertebrates on the streamside and whose toes are only partially webbed

Erithacus *eh-ri-THAK-us*
Robin, as in *Erithacus rubecula*, the European Robin

Erlangeri *er-LAN-ger-eye*
After Carol von Erlanger, German collector, as in *Calandrella erlangeri*, Erlanger's Lark

Erolia *eh-ROL-ee-a*
From *erolie*, a word made up by French ornithologist Vieillot, definition unclear, as in *Erolia* (now *Calidris*) *maritima*, the Purple Sandpiper

Erythrauchen *eh-ri-THRAW-ken*
Greek, *erythros*, red, and *auchen*, the neck, throat, as in *Accipiter erythrauchen*, the Rufous-necked Sparrowhawk

Erythrinus *eh-ri-THRY-nus*
Greek, *erythros*, red, and Latin *-inus*, pertaining to, as in *Carpodacus erythrinus*, the Common Rosefinch

Erythrocephala, -us *eh-rith-ro-se-FAL-a/us*
Greek, *erythros*, red, and Latin *cephala*, head, as in *Myzomela erythrocephala*, the Red-headed Myzomela

Erythrocercum, -us *eh-rith-ro-SIR-kum/kus*
Greek, *erythros*, red and *cerco*, tail, as in *Philydor erythrocercum*, the Rufous-rumped Foliage-gleaner

Erythrochlamys *eh-rith-ro-KLAM-is*
Greek, *erythros*, red, *chlamys*, cloak, as in *Calendulauda erythrochlamys*, the Dune Lark; some races have a rufous wash to the upper parts

Erythrocnemis *eh-rith-rok-NEM-is*
Greek, *erythros*, red, and *kneme*, leg, as in *Pomatorhinus erythrocnemis*, the Black-necklaced Scimitar Babbler, with rusty-colored thigh feathers

Erythrogaster, -trus *eh-rith-ro-GAS-ter/trus*
Greek, *erythros*, red, *gaster*, belly, as in *Laniarius erythrogaster*, the Black-headed Gonolek, with a bright chest and abdomen

Erythrogenys *eh-rith-ro-JEN-is*
Greek, *erythros*, red, and *genys*, jaw, as in *Psittacara erythrogenys*, the Red-masked Parakeet

Erythrogonys *eh-rith-ro-GON-is*
Greek, *erythros*, red, and *gony*, knee, as in *Erythrogonys cinctus*, the Red-kneed Dotterel

Pomatorhinus erythrocnemis, Black-necklaced Scimitar Babbler

ERITHACUS

There are many birds called robins—bush-robins, forest-robins, magpie-robins, and thrushes like the American Robin, *Turdus migratorius*, and Rufous-collared Robin, *T. rufitorques*, most with the familiar red breast, and the Flame Robin, *Petroica phoenicea*. But unlike *E. rubecula*, the European Robin, they are not actually in the genus *Erithacus* (eh-ri-THAK-us), a Latin word meaning robin and presumably referring to the European Robin. Once thought to be a thrush, it is now considered an Old World Flycatcher, specifically a chat. There are several myths and folktales explaining the red breast. One says it came from the blood of Christ when the bird pulled a thorn out of his crown. Another says that the bird saved a father and son on a cold night by fanning the flames of a fire with its wings. The bird's breast is orange-colored rather than red, but orange was not a known color until the sixteenth century.

The name robin comes from the fifteenth century and perhaps earlier, shortened from Robin Redbreast or Robin Goodfellow, but it was not applied to the European songbird until the mid-eighteenth century. Today the name applies to people, airplanes, ships, and fictional characters. There are only three species in the genus *Erithacus*, the other two being *E. akahige*, the Japanese Robin, and *E. komadori*, the Ryukyu Robin.

The European Robin ranges from northern Scandinavia to northern Africa; there are different populations that vary somewhat in coloration and are considered subspecies. The most different is the Canary Islands Robin, which has a white eye ring that the European population does not possess.

Erithacus rubecula, European Robin

The Japanese Robin, *E. akahige*, is found in China, Japan, Korea, Thailand, Vietnam, and Russia, and has an orange head rather than breast, and the Ryukyu Robin, *E. komadori*, with an orange crown, nape, back, and tail, is restricted to the Nansei Shoto archipelago of Japan.

All the *Erithacus* birds are woodland species, although the European Robin is common in British gardens where it follows gardeners tilling the soil in search of invertebrates. Being small birds these robins have a high mortality rate, especially when young, and have an average life span of just over one year.

Erithacus komadori, Ryukyu Robin

The Ryukyu Robin is found only in the Nansei Shoto archipelago between southern Japan and Taiwan, sometimes called the Galapagos of the Western Pacific.

Myzomela erythromelas, Black-bellied Myzomela

Erythroleuca eh-rith-ro-LOY-ka
Greek, *erythros*, red, and *leuca*, white, as in *Grallaria erythroleuca*, the Red-and-white Antpitta

Erythrolophus eh-rith-ro-LO-fus
Greek, *erythros*, red, and *lophus*, crest, as in *Tauraco erythrolophus*, the Red-crested Turaco

Erythromelas eh-rith-ro-MEL-as
Greek, *erythros*, red, and *melas*, black, as in *Myzomela erythromelas*, the Black-bellied Myzomela, with a red head

Erythronotos, -us, -a eh-rith-ro-NO-tos/tus/ta
Greek, *erythros*, red, and *noto*, back or south, as in *Estrilda erythronotos*, the Black-faced Waxbill. The species name is a misnomer as it is the lower abdomen and rump that are reddish

Erythrophthalma, -us eh-rith-ro-THAL-ma/mus
Greek, *erythros*, red, and *ophthalmos*, the eye, as in *Netta erythrophthalma*, the Southern Pochard; males have red eyes

Erythropleura eh-rith-ro-PLUR-a
Greek, *erythros*, red, and *pleura*, side, ribs, as in *Ptiloprora erythropleura*, the Rufous-sided Honeyeater

Erythrops eh-RI-throps
Greek, *erythros*, red and *ops*, the face, as in *Quelea erythrops*, the Red-headed Quelea

Erythroptera eh-rith-ROP-ter-a
Greek, *erythros*, red, and *pteron*, wing, as in *Prinia erythroptera*, the Red-winged Prinia

Erythropus eh-rith-RO-pus
Greek, *erythros*, red, and *pous*, foot, as in *Accipiter erythropus*, the Red-thighed Sparrowhawk

Erythropygia, -us eh-rith-ro-PIH-jee-a/us
Greek, *erythros*, red, and *puge*, rump, as in *Sturnia erythropygia*, the White-headed Starling

Erythrorhyncha, -chos eh-rith-ro-RIN-ka/kos
Greek, *erythros*, red, and Latin, *rhynchus*, bill, as in *Anas erythrorhyncha*, the Red-billed Teal

Erythrothorax eh-rith-ro-THOR-aks
Greek, *erythros*, red, and *thorax*, breast, as in *Dicaeum erythrothorax*, the Flame-breasted Flowerpecker

Erythrura, -us eh-rith-ROO-ra/rus
Greek, *erythros*, red, and *oura*, tail, as in *Erythrura viridifacies*, the Green-faced Parrotfinch

Estrilda es-TRIL-da
Derivation perhaps from the German *Wellenastrild*, the Waxbill, as in *Estrilda atricapilla*, the Black-headed Waxbill

Euchlorus you-KLOR-us
Greek *eu*, well or good, and *chlor-o*, green, as in *Passer euchlorus*, the Arabian Golden Sparrow (yellow with a greenish hue)

Eudocimus you-DOE-si-mus
Greek *eu*, well or good, and *docimus*, excellent, of good repute, as in *Eudocimus albus*, the American White Ibis, with a stately appearance

Eudromia you-DROM-ee-a
Greek, *eu*, well or good, and *dromos*, running, a race, as in *Eudromia elegans*, the Elegant Crested Tinamou, a good runner and poor flier

Eudynamys you-DY-na-mus
Greek, *eu*, well or good, and *dynam*, power, energy, as in *Eudynamys scolopaceus*, the Asian Koel; Koel is onomatopoeic

Eugenes you-JEN-eez
Greek, *eu*, well or good, and *genos*, birth, as in *Eugenes fulgens*, the Magnificent Hummingbird; refers to its size and magnificent coloration

Eugralla *you-GRAL-la*
Greek, *eu*, well or good, and Latin, *gralla*, stilts, as in *Eugralla paradoxa*, the Ochre-flanked Tapaculo, with long legs

Eulabeornis *you-la-be-OR-nis*
Greek, *eulab*, wary, cautious, and *ornis*, bird, as in *Eulabeornis castaneoventris*, the Chestnut Rail

Euleri *YOU-ler-eye*
After Carl Euler, Swiss Counsel in Rio de Janeiro, Brazil, as in *Coccyzus euleri*, the Pearly-breasted Cuckoo

Eulophotes *you-lo-FOE-teez*
Greek, *eu*, well or good, and Latin, *lophus*, crest, as in *Egretta eulophotes*, the Chinese Egret

Eumyias *you-MY-yas*
Greek, *eu*, well or good, and *muia*, fly, as in *Eumyias indigo*, the Indigo Flycatcher

Euodice *you-O-di-see*
Greek, *eu*, well or good, and *odi*, song, as in *Euodice cantans*, the African Silverbell

Euphagus *you-FAY-gus*
Greek, *eu*, well or good, and *phagein*, to eat, as in *Euphagus cyanocephalus*, Brewer's Blackbird, an omnivore

Eupherusa *you-fer-OO-sa*
Greek, *eu*, well or good, and *pher*, to bear, as in *Eupherusa nigriventris*, the Black-bellied Hummingbird, most likely in reference to its bearing or posture

Euphonia *you-FONE-ee-a*
Greek, *eu*, well or good, and *phon*, sound or voice, as in *Euphonia plumbea*, the Plumbeous Euphonia

Euplectes *you-PLEK-teez*
Greek, *eu*, well or good, and *plectos*, twisted or braided, as in *Euplectes afer*, the Yellow-crowned Bishop; the genus name refers to the bird's complex braided nest construction

Eupoda *you-PO-da*
Greek, *eu*, well or good, and *pous*, foot, as in *Eupoda* (now *Charadrius*) *montanus*, the Mountain Plover

Euptilotis *youp-til-O-tis*
Greek, *eu*, well or good, *ptilon*, feather, and *otis*, ear, as in *Euptilotis neoxenus*, the Eared Quetzal

Eurocephalus *you-ro-se-FAL-us*
Greek, *euro*, broad, and Latin, *cephala*, head, as in *Eurocephalus rueppelli*, the White-rumped Shrike, with a large head characteristic of shrikes

Europaea *you-ro-PEE-a*
Europe, as in *Sitta europaea*, the Eurasian Nuthatch

Euryceros *you-ri-SIR-os*
Greek, *euro*, broad, and *cera*, horn, as in *Euryceros prevostii*, the Helmet Vanga

Eurylaimus *you-ri-LIE-mus*
Greek, *euro*, broad, and *laimos*, throat, as in *Eurylaimus javanicus*, the Banded Broadbill

Eurynorhynchus *you-ri-no-RINK-us*
Greek, *euryno*, widen, and Latin, *rhynchus*, bill, as in *Eurynorhynchus pygmeus*, the Spoon-billed Sandpiper

Euryptila *you-rip-TIL-a*
Greek, *euro*, broad, and *ptila*, feather, as in *Euryptila subcinnamomea*, the Cinnamon-breasted Warbler

Eurypyga *you-ri-PI-ga*
Greek, *euro*, broad, and *puge*, rump, as in *Eurypyga helias*, the Sunbittern

Egretta eulophotes, Chinese Egret

LATIN IN ACTION

Newton's Parakeet is extinct. In 1872 a female was collected and two years later a male; these are the specimens from which the species was described and are the only ones in existence. The parakeet once inhabited the small island of Rodrigues, part of the Republic of Mauritius and located in the Indian Ocean about 350 kilometers east of Mauritius. Small islands are biologically fragile and extinction rates are much higher than on larger islands, but all islands are more ecologically unstable than the mainland. Perhaps the most famous island extinction is that of the Dodo, *Raphus cucullatus*, on Mauritius in 1690. Another parakeet, the Echo Parakeet, *Psittacula eques*, was represented by only three pairs in the 1980s; today, there are around 500. Lifting endemic species such as the Mauritius Kestrel, *Falco punctatus*, Pink Pigeon, *Nesoenas mayeri*, Rodrigues Warbler, *Acrocephalus rodericanus*, Rodrigues Fody, *Foudia flavicans*, and Echo Parakeet from the brink of extinction, Mauritius became the most successful country in the world at saving endangered species.

Psittacula exsul, Newton's Parakeet

Eurystomus *you-ri-STO-mus*
Greek, *euro*, broad, and *stomus*, mouth, as in *Eurystomus orientalis*, the Oriental Dollarbird, which has a broad bill

Everetti *EV-ver-et-tye*
After Alfred Everett, a British administrator and collector in the East Indies, as in *Rhyticeros everetti*, Sumba Hornbill

Eversmanni *EH-verz-man-nye*
After Alexander Eversmann, a Russian lepidopterist, as in *Columba eversmanni*, Yellow-eyed Pigeon

Ewingii *you-WING-ee-eye*
After Thomas Ewing, Australian teacher, naturalist, and collector, as in *Acanthiza ewingii*, the Tasmanian Thornbill

Excalfactoria *eks-kal-fak-TOR-ee-a*
Ex, out of, *cal*, heat, and *factoria*, place of production, because Chinese used these birds as hand-warmers, as in *Excalfactoria chinensis*, the King Quail

Excubitor *eks-KOO-bi-tor*
Sentinel, watchman, from *excubare*, out of doors, as in *Lanius excubitor*, the Northern or Great Gray Shrike, well known for watching from high vantage points

Exilis *eks-IL-is*
Small, delicate, as in *Psaltria exilis*, the Pygmy Bushtit

Eximia, -us, -um *ex-IM-ee-a/us/um*
Exceptional or uncommon, as in *Buthraupis eximia*, the Black-chested Mountain Tanager

Explorator *eks-PLOR-at-or*
Explorer, investigator, as in *Zosterops explorator*, the Fiji White-eye

Exsul *EKS-ool*
Exsula, stranger, exile, as in *Psittacula exsul*, the extinct Newton's Parakeet, endemic to the island of Rodrigues, Indian Ocean

Externa *eks-TURN-a*
Outside, external, as in *Pterodroma externa*, the Juan Fernandez Petrel. This bird breeds only on an island off the coast of Chile

Exustus *eks-US-tus*
Exust, burned up, consumed, as in *Pterocles exustus*, the Chestnut-bellied Sandgrouse; may refer to the fact that this bird lives in very hot dry, environments

F

Fabalis *fa-BAL-is*
Faba, bean, as in *Anser fabalis*, the Taiga Bean Goose, the common name probably from its habit of grazing in bean fields

Falcata, -us *fal-KA-ta/tus*
Falcis, sickle, as in *Anas falcata*, the Falcated Duck, with its sickle-shaped tertiary feathers

Falcinellus *fal-sin-EL-lus*
Falcis, sickle, as in *Limicola falcinellus*, the Broad-billed Sandpiper, perhaps for its downcurved bill tip

Falcipennis *fal-si-PEN-nis*
Falcis, sickle, and *penna*, feather, as in *Falcipennis falcipennis*, the Siberian Grouse, from its swept-back wings in flight

Falcirostris *fal-si-ROSS-tris*
Falcis, sickle, and *rostris*, beak, bill, as in *Sporophila falcirostris*, Temminck's Seedeater

Falco *FAL-ko*
Curved blade, sickle, as in *Falco concolor*, the Sooty Falcon, with a hooked beak

Falcularius *fal-koo-LAR-ee-us*
Falcis, sickle, and *-arius*, pertaining to, as in *Campylorhamphus falcularius*, the Black-billed Scythebill

Falculea *fal-KOOL-ee-a*
Falcis, sickle, as in *Falculea palliata*, the Sickle-billed Vanga

Falkensteini *FAL-ken-stine-eye*
After Johann Falkenstein, German surgeon and collector, as in *Chlorocichla falkensteini*, Falkenstein's Greenbul

Falklandicus *falk-LAND-ih-kus*
After the Falkland Islands, as in *Charadrius falklandicus*, the Two-banded Plover

Fallax *FAL-laks*
Fallac, deceptive, as in *Leucippus fallax*, the Buffy Hummingbird, deceptive probably because of its unusually dull color for a hummingbird

Sporophila falcirostris, Temminck's Seedeater

Familiaris, -e *fa-mil-ee-AR-is/-ee*
Familia, family, household, as in *Certhia familiaris*, the Eurasian Treecreeper

Famosa *fam-OS-a*
Fama, reputation, tradition, as in *Nectarinia famosa*, the Malachite Sunbird

Fanny, -i *FAN-nee/neye*
After Francis "Fanny" Wilson, wife of collector Edward Wilson, as in *Myrtis fannyi*, the Purple-collared Woodstar

Fasciata, -us *fas-ehe-AH-ta/tus*
Banded, as in *Atticora fasciata*, the White-banded Swallow

Fasciatoventris *fas-see-a-toe-VEN-tris*
Fascia, band, and *ventris*, belly, as in *Pheugopedius fasciatoventris*, the Black-bellied Wren

Fasciicauda *fas-see-eye-KAW-da*
Fascia, band, and *cauda*, tail, as in *Pipra fasciicauda*, the Band-tailed Manakin

Fasciinucha *fas-see-eye-NOO-ka*
Fascia, band, and *nucha*, nape, as in *Falco fasciinucha*, the Taita Falcon, of the Taita Hills of Kenya

FALCO

There are 37 species of birds comprising the genus *Falco* (*Fal-ko*), the falcons, from the Latin *falx*, curved blade, sickle. They may be so named for their talons, their curved beak, or the shape of their outspread wings. While hawks and falcons share some features, they are in different families: hawks and eagles are in Accipitridae and falcons in Falconidae. Falcons differ from hawks in that they are often smaller, with longer, thinner wings, and a tooth-like notch on the bill. Falcons typically catch their prey in mid-air while hawks tend to snatch up their food items from the ground. The Peregrine Falcon (*F. peregrinus*, Latin for wanderer) is reputed to be able to exceed 185 miles (300 kilometers) per hour in a dive. Falcons are widely distributed, but the Peregrine is the most widespread, found almost everywhere between the polar regions except high mountains, deserts, and tropical areas. The Barbary Falcon, looking a lot like the Peregrine, is named *F. pelegrinoides*, Latin *pelegrinus*, meaning Peregrine Falcon, and the Greek suffix *-oides*, meaning resembling.

Kestrels are another *Falco* subgroup. They are smaller than the peregrine group, and, unlike most falcons, sexually dimorphic. Relatively colorful, they tend to hover and dive on their bird or mammal prey rather than catch it mid-air.

Falco peregrinus, Peregrine Falcon

Kestrel comes from the French *crécerelle*, rattle, apparently from their call. The American Kestrel (*F. sparverius*, Latin for sparrowhawk) ranges throughout the Americas from Alaska to Tierra del Fuego.

Then there are the hobbies, between the peregrine group and kestrels in size and dark gray in color. More aerial than the kestrels, they take small birds and large insects in flight. The Eurasian Hobby is found throughout that region. Its scientific designation is *F. subbuteo* (Latin *sub*, near to, and *buteo*, buzzard) and hobby comes from the Old French *hobet*, meaning falcon and referring to its flight—up and down (like a hobby horse).

As with most raptors (a bird of prey, such as hawks and owls) females are usually larger than the males. They lay eggs that hatch asynchronously because incubation usually begins after the first egg is laid, so the chicks are of different sizes as they develop. In times of food scarcity, the first-born chicks survive because they are bigger and better at begging for food.

Falco amurensis, Amur Falcon

The Amur Falcon has a round-trip migratory journey of 14,000 miles from southern Africa to Asia each year.

Fasciiventer *fas-see-eye-VEN-ter*
Fascia, band, and *ventris*, belly, as in *Melaniparus fasciiventer*, the Stripe-breasted Tit

Fasciogularis *fas-see-o-goo-LAR-is*
Fascia, band, and *gularis*, throat, as in *Gavicalis fasciogularis*, the Mangrove Honeyeater

Fasciolata, -us *fas-see-o-LAT-a/us*
Fasciat-, banded, as in *Crax fasciolata*, the Bare-faced Curassow

Fastuosa *fas-to-O-sa*
Fastuosus, proud, haughty, as in *Tangara fastuosa*, the Seven-colored Tanager, aptly named for its spectacular plumage

Feae *FAY-ee*
After Leonardo Fea, Italian naturalist, as in *Turdus feae*, the Gray-sided Thrush

Featherstoni *FE-ther-stone-eye*
After Earl Featherston, a superintendent of the province of Wellington, New Zealand, as in *Phalacrocorax featherstoni*, the Pitt Shag or Featherstone's Shag

Fedoa *fe-DOE-a*
After an old name for the godwit, as in *Limosa fedoa*, the Marbled Godwit

Felix *FEE-liks*
Happy, blessed, fertile, as in *Pheugopedius felix*, the Happy Wren, probably from its song

Femoralis *fe-mor-AH-lis*
Femur, thigh, as in *Falco femoralis*, the Aplomado Falcon, with reddish thighs, distinct from the rest of its plumage

Ferina *fe-REEN-a*
Game, flesh of wild animals, as in *Aythya ferina*, the Common Pochard, probably so named because it once was commonly eaten

Ferminia *fair-MIN-ee-a*
After Fermin Cervera, Spanish soldier and naturalist, as in *Ferminia cerverai*, the Zapata Wren

Fernandensis *fer-nan-DEN-sis*
After the Juan Fernández Islands off Chile, as in *Sephanoides fernandensis*, the Juan Fernandez Firecrown

Ferox *FER-oks*
Fierce, as in *Myiarchus ferox*, the Short-crested Flycatcher

Ferreorostris *fer-ree-o-ROSS-tris*
Ferro, iron, and *rostris*, bill, as in *Carpodacus ferreorostris*, the Bonin Grosbeak

Ferreus *FER-ree-us*
Ferro, iron, as in *Saxicola ferreus*, the Gray Bush Chat; refers to the male bird's iron-colored plumage

Ferruginea, -us *fer-roo-JIN-ee-a/us*
Rust-colored, as in *Muscicapa ferruginea*, the Ferruginous Flycatcher

Ferrugineifrons *fer-roo-jin-ee-EYE-fronz*
Ferrugineus, rust-colored, and *frons*, forehead, as in *Bolborhynchus ferrugineifrons*, the Rufous-fronted Parakeet

Ferrugineipectus *fer-roo-jin-ee-eye-PEK-tus*
Ferrugineus, rust-colored, and *pectus*, breast, as in *Grallaricula ferrugineipectus*, the Rusty-breasted Antpitta

Ferrugineiventre *fer-roo-jin-ee-eye-VEN-tree*
Ferrugineus, rust-colored, and *ventr*, belly, as in *Conirostrum ferrugineiventre*, the White-browed Conebill

Festiva *fes-TEE-va*
Festive, holiday mood, as in *Amazona festiva*, the Festive Amazon

Ficedula *fee-se-DOO-la*
Small bird, fig-pecker, as in *Ficedula hypoleuca*, the European Pied Flycatcher

Pheugopedius felix, Happy Wren

Coracina fimbriata,
Lesser Cuckooshrike

Figulus *fi-GOO-lus*
Potter, earthenware maker, from *fingere*, transform, as in *Furnarius figulus*, the Band-tailed Hornero, that builds oven-shaped nests. Hornero comes from the Spanish *horno*, meaning oven

Filicauda *fi-li-KAW-da*
Fili, thread, and *cauda*, tail, as in *Pipra filicauda*, the Wire-tailed Manakin

Fimbriata, -um *fim-bree-AH-ta/tum*
Fimbri-, fringe, fibers, as in *Coracina fimbriata*, the Lesser Cuckooshrike, with white fringes on its flight feathers

Finschi, -ii *FINCH-eye/ee-eye*
After Friedrich Finsch, a German ethnographer and naturalist, as in *Scleroptila finschi*, Finsch's Francolin

Fischeri *FISH-er-eye*
After Gustav Fischer, German explorer, as in *Agapornis fischeri*, Fischer's Lovebird

Fistulator *fiss-too-LA-tor*
Fistulare, one who plays the reed-pipe, as in *Ceratogymna* (now *Bycanistes*) *fistulator*, the Piping Hornbill, probably describing its call

Flabelliformis *fla-bel-li-FORM-is*
Flabellum, small fan, and *form-*, shape, as in *Cacomantis flabelliformis*, the Fan-tailed Cuckoo

Flagrans *FLAY-granz*
Burning, blazing, fiery, as in *Aethopyga flagrans*, the Flaming Sunbird

Flammea, -us, -olus *FLAM-me-a/us/FLAM-me-o-lus*
Flamme-, flame-colored, as in *Acanthis flammea*, the Common Redpoll

Flammiceps *FLAM-mi-seps*
Flammeus, flame-colored, and *ceps*, head, as in *Cephalopyrus flammiceps*, the Fire-capped Tit; *flammiceps* redundant as *Cephalopyrus* also means flame-colored head

Flammigerus *flam-mi-JER-us*
Flammeus, flame-colored, and *gero-*, to carry, bear, as in *Ramphocelus flammigerus*, the Flame-rumped Tanager

Flammulatus, -a *flam-moo-LA-tus/ta*
Little flame, as in *Megabyas flammulatus*, the African Shrike-flycatcher

Flava *FLA-va*
Flavus, yellow, as in *Motacilla flava*, the Western Yellow Wagtail

Flavala *fla-VAL-a*
Flavus, yellow, and *ala*, wing, as in *Hemixos flavala*, the Ashy Bulbul

Flaveola, -lus *flav-ee-O-la/lus*
Flavus, yellow, as in *Coereba flaveola*, the Bananaquit

Acanthis flammea,
Common Redpoll

Flavescens *FLAV-es-senz*
Flavescere, becoming golden, yellow, as in *Celeus flavescens*, the Blond-crested Woodpecker

Flavicans *FLAV-ih-kanz*
Flavere, being golden or yellow-colored, as in *Prinia flavicans*, the Black-chested Prinia, with yellow underparts

Flavicapilla *flav-ih-ka-PIL-la*
Flavus, yellow, and *capilla*, hair, as in *Xenopipo flavicapilla*, the Yellow-headed Manakin

Flaviceps *FLAV-ih-seps*
Flavus, yellow, and *ceps*, head, as in *Auriparus flaviceps*, the Verdin, with a yellow face and head

Flavicollis *flav-ih-KOL-lis*
Flavus, yellow, and *collis*, neck, as in *Yuhina flavicollis*, the Whiskered Yuhina, with both an orange and yellow collar

Flavifrons *FLAV-ih-fronz*
Flavus, yellow, and *frons*, forehead, as in *Melanerpes flavifrons*, the Yellow-fronted Woodpecker

Flavigaster, -ogaster *flav-ih-GAS-ter/flav-o-GAS-ter*
Flavus, yellow, and *gaster*, stomach, as in *Hyliota flavigaster*, the Yellow-bellied Hyliota

Flavigula, -aris *flav-ih-GOO-la/flav-ih-goo-LAR-is*
Flavus, yellow, and *gula*, throat, as in *Crithagra flavigula*, the Yellow-throated Seedeater

Flavinucha *flav-ih-NOO-ka*
Flavus, yellow, and *nucha*, nape, as in *Chrysophlegma flavinucha*, the Greater Yellownape

Flavipennis *flav-ih-PEN-nis*
Flavus, yellow, and *pennis*, feather or wing, as in *Chloropsis flavipennis*, the Philippine Leafbird. Although mostly green, there are yellow fringes on the primary feathers

Flavipes *flav-IP-eez*
Flavus, yellow, and *pes*, foot, as in *Tringa flavipes*, the Lesser Yellowlegs

Flaviprymna *fla-vi-PRIM-na*
Flavus, yellow, and Greek, *prumnos*, rear end, as in *Lonchura flaviprymna*, the Yellow-rumped Mannikin

LATIN IN ACTION

There are over 200 species of woodpecker, found all over the world except Australia, New Guinea, and Madagascar. They belong to the family Picidae, which also includes sapsuckers, wrynecks, piculets, flamebacks, and flickers. The Greater Yellownape, *Chrysophlegma flavinucha*, has a very large range across Asia. In Roman mythology, Picus was a handsome king but when the witch Circe tried to seduce him she turned him into a woodpecker for his apparent ability to interpret bird omens. Woodpeckers have zygodactyl (Greek for yoke-finger) feet, meaning two toes face forward and two backward, in a sort of X pattern, and their tail feathers are stiff. These two adaptations allow the woodpecker to lean back and prop itself against a tree while it pounds away at the bark. Special adaptations of the head and neck prevent injury to the bird while it uncovers food under tree bark, excavates nests holes, or even hammers on metal posts in order to announce its territory.

Chrysophlegma flavinucha, Greater Yellownape

CHRISTIAN JOUANIN
(B. 1925)

Christian Jouanin, born in Paris in 1925, is a well-known and respected ornithologist whose specialty is petrels. At the age of 15 he began work at the National Museum of Natural History under the supervision of Jacques Berlioz, the head of the Department of Ornithology at the museum. Later he worked with Jean Dorst, president of the 16th International Ornithological Congress, who followed Berlioz as the head of the department. Jouanin and Dorst wrote the species description for the rare Djibouti Francolin, *Pternistis ochropectus*, whose type specimen they brought to the museum. Its specific epithet comes from the Greek, *ochros*, the color ocher, and the Latin, *pectus*, breast. Today there is some question about the species status of this bird because it is very closely related to other francolins and intermediate in both location and anatomy to other species.

In 1955 Jouanin published his first species description. Studying the Mascarene Petrel, *Pseudobulweria aterrima*, he found that this species was actually two species. He described and split off what is now named Jouanin's Petrel, *Bulweria fallax*. Both species are very rare and critically endangered. Continuing work on the Mascarene Petrel, which may be very close to extinction, he discovered yet another new species, Barau's Petrel, which he found breeding in the French territory of Réunion Island in the Indian Ocean. The name Barau is after Armand Barau, an agricultural engineer and ornithologist from Réunion. This petrel is one of the most recently discovered species of seabird, not described until 1964, although it had been long known to the local inhabitants of the island. Jouanin also discovered differences in the populations of Audubon's Shearwaters in the Seychelles and Réunion, and designated them a subspecies. He named the Seychelles subspecies after his wife Nicole, *Puffinus bailloni nicolae*.

Jouanin spent many years in the Indian Ocean, but in 1963 he launched studies on seabirds of the Atlantic Ocean when he joined Francis Roux on an expedition to the Savage Islands. Then he went on to collect specimens and data on Cory's Shearwater, *Calonectris borealis*, in the waters around Madeira with Alex Zino, a Portuguese ornithologist; Zino's Petrel, *Pterodroma madeira*, is named for this colleague.

Diomedea exulans,
Wandering Albatross

At up to 11 foot 10 inches, the wingspan of the Wandering Albatross is the longest of any bird. This albatross can stay aloft for days by taking advantage of wind and wave currents close to the ocean's surface.

Jouanin is a recognized expert on the Order Procellariformes, a group of seabirds in four families: albatrosses, storm petrels, diving petrels, petrels, and shearwaters. These birds, almost exclusively pelagic and found across the world's oceans, are often called tubenoses because of their tube-shaped nasal passages. Jouanin co-authored the chapter on Procellariformes with J. L. Mougin in *Peters Check-list of the Birds of the World*.

He was founder and director of the MAR Bureau (MAR stands for the first three letters of the word for wetlands in English, MARshes, French, *MARecages*, and Spanish, *MARismas*), an organization involved with the preservation of wetlands; general secretary of the French Société National de Protection de la Nature; vice-president of the International Union for the Conservation of Nature; and member of the Permanent Executive Committee of the International Ornithological Committee.

Calonectris borealis,
Cory's Shearwater

Cory's Shearwater was named after Charles Cory, who collected 19,000 bird specimens and eventually became curator of ornithology at the Field Museum of Chicago.

Hydrobates pelagicus,
European Storm Petrel

"[*Oiseaux De La Réunion* by Barre, Barau, and Jouanin] should be mandatory for all school children of Réunion taking courses in geography on this magnificent island."

Francois Vuilleumier, Wilson Bulletin (June 1999)

Flavirictus *flav-ih-RIK-tus*
Flavus, yellow, and *rictus*, jaw, open mouth, as in *Meliphaga flavirictus*, the Yellow-gaped Honeyeater

Flavirostris, -a *flav-ih-ROSS-tris/tra*
Flavus, yellow, and *rostris*, bill, as in *Monasa flavirostris*, the Yellow-billed Nunbird

Flaviventer, -tris *flav-ih-VEN-ter/tris*
Flavus, yellow, and *venter*, underside, belly, as in *Dacnis flaviventer*, the Yellow-bellied Dacnis

Flavivertex *flav-ih-VER-teks*
Flavus, yellow, and *vertex*, highest point, as in *Myiopagis flavivertex*, the Yellow-crowned Elaenia, the common name from Greek *eleia*, olive, referring to its color

Flavovirens, -viridis, -virescens *flav-o-VIR-enz/flav-o-vir-ID-is/flav-o-vir-ES-sens*
Flavus, yellow, and *virere*, to be green, as in *Chlorospingus flavovirens*, the Yellow-green Bush Tanager

Flavus *FLA-vus*
Yellow, as in *Zosterops flavus*, the greenish yellow Javan White-eye

Floccosus *flok-KO-sus*
Flocc-, a lock of wool, flake, as in *Pycnoptilus floccosus*, the Pilotbird, whose species epithet may describe its loose plumage; its common name comes from its habit of following lyrebirds, taking prey that they flush

Floriceps *FLOR-ih-seps*
Flor-, flower, and *ceps*, head, as in *Anthocephala floriceps*, the Blossomcrown

Florida *flo-REE-da*
Floridis, flowering, as in *Tangara florida*, the Emerald Tanager

Floris *FLO-ris*
From Flores, an island in Indonesia, as in *Treron floris*, the Flores Green Pigeon

Florisuga *flor-ih-SOO-ga*
Flor, flower, and *sugere*, to suck, as in *Florisuga fusca*, the Black Jacobin, a nectar-feeding hummingbird

Fluviatilis *floo-vee-a-TIL-is*
Fluvialis, of a river, as in *Locustella fluviatilis*, the River Warbler

Fluvicola *floo-vi-KO-la*
Fluvi, river, and *cola*, dwell, as in *Fluvicola pica*, the Pied Water Tyrant

Foersteri *FUR-ster-eye*
After F. Földersteri, German botanist and collector, as in *Melidectes foersteri*, the Huon Melidectes, Huon after a New Guinea peninsula

Forbesi *FORBS-eye*
After Henry Forbes, a Scottish explorer and collector, as in *Rallicula forbesi*, Forbes's Forest Rail; also William Forbes, British anatomist, collector, and zoologist, as in *Charadrius forbesi*, Forbes's Plover

Forficatus, -a *for-fi-KA-tus/ta*
Forficata, forked, as in *Dicrurus forficatus*, the Crested Drongo, with a forked tail

Formicarius *form-ih-KAR-ee-us*
Of the ant, as in *Formicarius moniliger*, the Mayan Antthrush

Formicivora, -ous *form-ih-SI-vor-a/us*
Formica, an ant, and *vora*, eat, devour, as in *Formicivora grisea*, the Southern White-fringed Antwren

Formosa, -sus *for-MO-sa/sus*
Formosus, beautiful, as in *Sitta formosa*, the Beautiful Nuthatch

Formosae *for-MO-see*
After Formosa, now Taiwan, as in *Treron formosae*, the Whistling Green Pigeon

Florisuga fusca, Black Jacobin

Sitta formosa,
Beautiful Nuthatch

Forsteni *FOR-sten-eye*
After Eltio Forsten, Dutch botanist and collector, as in *Ducula forsteni*, the White-bellied Imperial Pigeon

Forsteri *FOR-ster-eye*
After Johann Forster, German clergyman and naturalist, as in *Sterna forsteri*, Forster's Tern

Fortis *FOR-tis*
Strong, powerful, as in *Geospiza fortis*, the Medium Ground Finch of the Galapagos

Foudia *FOO-dee-a*
Mascarene name for the Fody, as in *Foudia madagascariensis*, the Red Fody

Francesiae *fran-SES-ee-ee*
After Henrietta Frances Cole, patroness of science, as in *Accipiter francesiae*, Frances's Sparrowhawk

Francolinus *frank-o-LEEN-us*
Latinization of the Italian *francolino*, little hen, as in *Francolinus francolinus*, the Black Francolin

Frantzii *FRANTZ-ee-eye*
After Alexander von Franzius, German naturalist and collector, as in *Catharus frantzii*, the Ruddy-capped Nightingale-Thrush

Fraseri, -a, *FRAZ-er-eye/a*
After Louis Fraser, British zoologist and collector, as in *Deleornis fraseri*, Fraser's Sunbird

Frater *FRA-ter*
Brother, cousin, as in *Monarcha frater*, the Black-winged Monarch, apparently because of its gregarious habits

---

LATIN IN ACTION

The colorful beak of the Atlantic Puffin, *Fratercula arctica*, has caused it to be called a "sea parrot." The name puffin originally meant "fatling" which actually referred to the chicks of the Manx Shearwater (*Puffinus puffinus*). *Fratercula* means little brother, or little friar, referring to the way their feet are held together in flight as if they were praying. Puffins eat mostly small fish and their strong beaks, spiny upper palates, and raspy tongues allow them to carry on average 10 fish at a time. One was recorded carrying 62 fish at once!

Fratercula *fra-ter-KOO-la*
Frater, brother, and *-cula*, small, as in *Fratercula arctica*, the Atlantic Puffin (see box)

Fregata *fre-GA-ta*
From Middle French, *frigate*, a small, fast ship, as in *Fregata magnificens*, the Magnificent Frigatebird, with large, sail-like wings

Fregetta *fre-GET-ta*
Latinized form of the English frigate, a small ship, as in *Fregetta grallaria*, the White-bellied Storm Petrel

Fratercula arctica,
Atlantic Puffin

Frenatus, -a *fre-NA-tus/ta*
From *frenare*, to hold, curb, restrain, as in *Bolemoreus frenatus*, the Bridled Honeyeater, a reference to the face pattern, as if wearing a bridle

Fringilla, -aris, -arius, -inus *frin-JIL-la/ frin-jil-LAR-is/ee-us/frin-jil-EYE-nus*
Fringilla, finch, as in *Fringilla coelebs*, the Common Chaffinch, the common name from the Old English *ceaffinc*, literally chaff finch, because of its habit of eating chaff, waste grain

Fringilloides *frin-jil-LOY-deez*
Fringilla, finch, and Greek, *oides*, resembling, as in *Dolospingus fringilloides*, the White-naped Seedeater

Frontales *fron-TAL-eez*
Frons, the forehead, brow, as in *Cinclidium frontale*, the Blue-fronted Robin

Frontalis *fron-TAL-is*
Frons, forehead, brow, as in *Anarhynchus frontalis*, the Wrybill

Frontata, -us *fron-TAT-a/us*
Frons, the forehead, brow, as in *Tricholaema frontata*, the Miombo Pied Barbet, after the Miombo woodlands of Africa

Frugivorus *froo-ji-VOR-us*
Frugi, fruit, and *vora*, to eat, as in *Calyptophilus frugivorus*, the Eastern Chat-Tanager

Fucata *foo-KA-ta*
Fucare, color, paint, dye, as in *Emberiza fucata*, the Chestnut-eared Bunting

Fuciphagus *foo-si-FAY-gus*
Fuci, seaweed, and *phagus*, eater of, as in *Aerodramus fuciphagus*, the Edible-nest Swiftlet; the species name derives from a Chinese story of the birds swooping down into the ocean to collect material for their nests, actually made almost exclusively of saliva

Fuelleborni *FUL-le-born-eye*
After Friederich Fülleborn, German physician, as in *Laniarius fuelleborni*, Fülleborn's Boubou, common name from its call

Fuertesi *foo-EHR-tess-eye*
After Louis Agassiz Fuertes, ornithologist and bird artist, as in *Hapalopsittaca fuertesi*, Fuertes's Parrot

Fulgens *FUL-jenz*
Glittering, as in *Eugenes fulgens*, the Magnificent Hummingbird

Emberiza fucata, Chestnut-eared Bunting

Fulgidus *ful-JEE-dus*
Shining, gleaming, glittering, from *fulgere*, to flash or shine, as in *Caridonax fulgidus*, the Glittering Kingfisher

Fulica *ful-ee-ka*
Waterfowl, as in *Fulica americana*, the American Coot

Fulicarius *ful-ih-KAR-ee-us*
Coot-like, as in *Phalaropus fulicarius*, the Red or Gray Phalarope, which swims on the surface of water

Fuliginosa, -sus *ful-ih-ji-NO-sa/sus*
Fuligo, soot, and *os-a*, full of, as in *Strepera fuliginosa*, the Black Currawong, a native Australian name perhaps from the call

Fuligiventer *ful-ih-ji-VEN-ter*
Fuligo, soot, and *venter*, belly, as in *Phylloscopus fuligiventer*, the Smoky Warbler

Fuligula *ful-ih-GOO-la*
Fuligo, soot, and *gula*, throat, as in *Aythya fuligula*, the Tufted Duck, referring to the bird's predominantly black color

Fulmarus *ful-MAR-us*
Old Norse, *full*, foul, and *mar*, gull, as in *Fulmarus glacialis*, the Northern Fulmar; the genus name comes from the bird's habit of regurgitating a foul-smelling liquid when disturbed and its superficial similarity to gulls

Atlapetes fulviceps,
Fulvous-headed Brush Finch

Fulva FUL-va
Fulvus, brownish, as in *Pluvialis fulva*, the Pacific Golden Plover

Fulvescens *ful-VES-senz*
Fulvus, brownish, as in *Prunella fulvescens*, the Brown Accentor

Fulvicapilla *ful-vi-ka-PIL-la*
Fulvus, brownish, and *capilla*, hair, as in *Cisticola fulvicapilla*, the Piping Cisticola, with a rufous cap

Fulvicauda *ful-vi-KAW-da*
Fulvus, brownish, and *cauda*, tail, as in *Basileuterus fulvicauda*, the Buff-rumped Warbler

Fulviceps *FUL-vi-seps*
Fulvus, brownish, and *ceps*, head, as in *Atlapetes fulviceps*, the Fulvous-headed Brush Finch

Fulvicollis *ful-vi-KOL-lis*
Fulvus, brownish, and *collis*, neck, as in *Treron fulvicollis*, the Cinnamon-headed Green Pigeon

Fulvifrons *FUL-vi-fronz*
Fulvus, brownish, and *frons*, forehead, as in *Empidonax fulvifrons*, the Buff-breasted Flycatcher

Fulvigula *ful-vi-GOO-la*
Fulvus, brownish, and *gula*, throat, as in *Anas fulvigula*, the Mottled Duck

Fumigatus *foo-mi-GAT-us*
To smoke, as in *Contopus fumigatus*, the Smoke-colored Pewee

Funebris *foo-NE-bris*
Funeral, deadly, fatal, as in *Todiramphus funebris*, the Sombre or Funereal Kingfisher, pertaining to the bird's dark plumage

Funerea, -us *foo-NER-ee-a/us*
Deadly, funereal, as in *Drepanis funerea*, the extinct Black Mamo

Furcata, -tus *fur-KA-ta/tus*
Forked, as in *Tachornis furcata*, the Pygmy Palm Swift, with a forked tail

Fusca *FUSS-ka*
Fuscus, dark, dusky, as in *Gerygone fusca*, the Western Gerygone, pronounced *jer-IH-gon-ee*

Fuscata, -us *fuss-KA-ta/tus*
Fuscus, dark, as in *Lonchura fuscata*, the Timor Sparrow

Fuscescens *fuss-SES-senz*
Fuscus, dark, as in *Catharus fuscescens*, the Veery, although the bird is more of a brownish-red coloration

Fuscicauda *foo-shi-CAW-da*
Fuscus, dark, and *cauda*, tail, as in *Habia fuscicauda*, the Red-throated Ant Tanager

Fuscicollis *foo-shi-KOL-lis*
Fuscus, dark, dusky, and *collis*, neck, as in *Poicephalus fuscicollis*, the Brown-necked Parrot

Fuscirostris *foo-shi-ROSS-tris*
Fuscus, dark, dusky, and *rostris*, bill, as in *Talegalla fuscirostris*, the Black-billed Brushturkey

Fuscus *FUS-kus*
Dark, dusky, as in *Casiornis fuscus*, the Ash-throated Casiornis

Catharus fuscescens,
Veery

G

Gabela *ga-BEL-a*
After Gabela, in Angola, as in *Prionops gabela*, the Gabela Helmetshrike

Gabonensis *ga-bo-NEN-sis*
After Gabon, as in *Dendropicos gabonensis*, the Gabon Woodpecker

Gaimardi *gy-MAR-dye*
After Joseph Gaimard, French surgeon, explorer, and naturalist, as in *Phalacrocorax gaimardi*, the Red-legged Cormorant

Galactotes *ga-lak-TOT-eez*
Greek, *galaktos*, milk, and *otes*, resembling, as in *Erythropygia galactotes*, the Rufous-tailed Scrub Robin

Galapagoensis *ga-la-pa-go-EN-sis*
From the Galapagos, as in *Buteo galapagoensis*, the Galapagos Hawk

Galatea *ga-la-TEE-a*
After Galatea, a mythical Greek sea nymph, as in *Tanysiptera galatea*, the Common Paradise Kingfisher

Galbula *gal-BOO-la*
Galbulus, oriole, as in *Galbula galbula*, the Green-tailed Jacamar, jacamar from the Tupi language of South America

Galeata, -us *gal-ee-AT-a/us*
Helmeted, as in *Myiagra galeata*, the Slaty Monarch or Moluccan Flycatcher, helmet probably referring to the slight crest of most flycatchers

Galericulata *ga-ler-ih-koo-LA-ta*
Galer, cap, and *cul*, little, as in *Aix galericulata*, the Mandarin Duck, with a swept-back head crest

Galerita *gal-er-EE-ta*
Galer, cap, and *-ita*, little, as in *Cacatua galerita*, the Sulphur-crested Cockatoo

Galgulus *gal-GOO-lus*
Galbulus, oriole, as in *Loriculus galgulus*, the Blue-crowned Hanging Parrot. An oriole hangs its nest; some confusion caused the name to be attributed to this parrot, which hangs upsidedown from branches when seeking food

Buteo galapagoensis, Galapagos Hawk

Galinieri *gal-in-ee-AIR-eye*
After Joseph Galinier, French explorer of Abyssinia (now Ethiopia), as in *Parophasma galinieri*, the Abyssinian Catbird

Gallicolumba *gal-li-ko-LUM-ba*
Gallus, cock, and *columba*, dove, as in *Gallicolumba rufigula*, the Cinnamon Ground Dove

Gallinago *gal-li-NA-go*
Gallina, hen, and *gallus*, cock, chicken, as in *Gallinago gallinago*, the Common Snipe; implies that the bird resembled a hen to the namer

Gallinula *gal-li-NOO-la*
Little hen, the diminutive of *gallina*, as in *Gallinula chloropus*, the Common Moorhen

Gallirallus *gal-li-RAL-lus*
Galli, chicken, and *rallus*, rail or thin, as in *Gallirallus torquatus*, the Barred Rail. The phrase "thin as a rail" refers to the laterally flattened bodies of these birds, not railroad tracks

Gallopavo *gal-lo-PA-vo*
Galli, chicken, and *pavus*, peacock, as in *Meleagris gallopavo*, the Wild Turkey

Galloperdix *gal-lo-PER-diks*
Galli, chicken, and *perdix*, partridge, as in *Galloperdix spadicea*, the Red Spurfowl

Gallus *GAL-lus*
Galli, chicken, as in *Gallus gallus*, the Red Junglefowl; *Gallus gallus domesticus* is the familiar domesticated chicken

Gambeli, -ii *GAM-bel-eye/gam-BEL-ee-eye*
After William Gambel, American naturalist and collector, as in *Poecile gambeli*, the Mountain Chickadee and *Callipepla gambelii*, Gambel's Quail

Gambensis *gam-BEN-sis*
After Gambia, as in *Plectropterus gambensis*, the Spur-winged Goose

Gampsonyx *gamp-SON-iks*
Greek, *gampso*, curved, and *onux*, nail, claw, as in *Gampsonyx swainsonii*, the Pearl Kite

Plectropterus gambensis, Spur-winged Goose

Gampsorhynchus *gamp-so-RINK-us*
Greek, *gampso*, curved, and Latin, *rhynchus*, bill, as in *Gampsorhynchus torquatus*, the Collared Babbler with a somewhat hooked upper bill

Garleppi *GAR-lep-pye*
After Gustav Garlepp, a German collector, as in *Compospiza garleppi*, the Cochabomba Mountain Finch

Garrula, -us *gar-ROO-la/lus*
Chattering, as in *Ortalis garrula*, the Chestnut-winged Chachalaca

Garrulax *gar-ROO-laks*
Greek form of the Latin *garrulus*, talkative, chattering, as in *Garrulax canorus*, the Chinese Hwamei or Melodious Laughing Thrush

Garzetta *gar-ZET-ta*
Italian for egret, as in *Egretta garzetta*, the Little Egret

Gaudichaud *GAW-di-show-d*
After Charles Gaudichaud-Beupré, French pharmacist and explorer, as in *Dacelo gaudichaud*, the Rufous-bellied Kookaburra

Gavia *GAV-ee-a*
Seabird, loon, as in *Gavia stellata*, the Red-throated Loon or Diver

Gayi *GAY-eye*
After Claude Gay, French zoologist and collector, as in *Attagis gayi*, the Rufous-bellied Seedsnipe

Geelvinkiana, -um *gel-vink-ee-AN-a/um*
After a Dutch ship and family, as in *Micropsitta geelvinkiana*, the Geelvink Pygmy Parrot

Gelochelidon *je-lo-KEL-ih-don*
Greek, *gelo*, laugh, *chelidon*, swallow, as in *Gelochelidon nilotica*, the Gull-billed Tern, which feeds mainly on insects in flight (as do swallows) and has a distinct laughing call

Genei *JEN-nay-eye*
After Guiseppe Gené, Italian naturalist, as in *Chroicocephalus genei*, the Slender-billed Gull

Genibarbis *jen-ih-BAR-bis*
Gena, the cheek or chin, and *barbus*, barbel or whisker, as in *Myadestes genibarbis*, the Rufous-throated Solitaire

Gavia

Most people are familiar with loons, or divers, members of the genus *Gavia* (*GAV-ee-a*), such as the Common Loon or Diver, *G. immer*. *Gavia* is the Latin word for seabird, originally used to describe a sea duck. The word loon, sometimes associated with the moon, as in lunar, actually comes from the Norwegian term *lom* or *lum*, meaning clumsy. Loons are clumsy on land because their webbed feet are located far back on the body; very efficient for swimming but terrible for walking on land. The name might also have come from the Dutch *loen*, a crazy person. Loon, loony, and lunatic have all come to mean crazy, as in "crazy as a loon." The tremolo, the best known of their eerie calls, sounds like crazy laughter, hence the application of the name to an unbalanced person.

Loons, in their own order, Gaviiformes, and family, Gaviidae, are found only in North America and Eurasia. In Europe they are commonly called divers because they dive for their food, which they occasionally spear with their pointed bill. Most of their prey consists of fish, but frogs and crayfish are also in their diet. Since they feed by sight, they are found only on clear lakes. They can dive as deep as 200 feet (60 meters) in pursuit of prey, not only because of their rearward, laterally flattened legs and webbed feet, but also because their bones, unlike the hollow bones of most birds, are solid. In addition, they can flatten their feathers to expel air bubbles and even adjust their buoyancy so only their head is above water. For digesting their food, they ingest small stones called gastroliths to help grind food in their stomach. Loons are heavy, bulky birds, up to 13 pounds (6 kilograms), and require a long run across the water for takeoff. They would rather dive than fly to escape predators.

Gavia immer,
Great Northern Loon or Diver

There are four (or five, according to some) species of loons/divers, all of which nest in freshwater lakes in northern North America and Eurasia. After breeding, they move to coastal waters of the Atlantic or Pacific to spend the winter. In the late winter or early spring, most loons shed all of their feathers in a short period of time, making them flightless for several weeks until the new flight feathers regrow.

The top figure is the Great Northern Loon or Diver, *Gavia immer*, the most abundant and widespread of North American loons. In the middle left is the Red-throated Loon, *Gavia stellata*; in the middle right the Yellow-billed Loon, *Gavia adamsii*; and on the bottom is the Pacific Loon, *Gavia pacifica*, which is nearly identical to the much rarer and not pictured Black-throated Loon, *Gavia arctica*.

Geocolaptes olivaceus, Ground Woodpecker

Gentilis *jen-TIL-is*
Of the same family or clan, as in *Accipiter gentilis*, the Northern Goshawk. Common name comes from the Old English *gōsheafoc*, goose-hawk

Geobates *jee-o-BAT-eez*
Greek, *geo*, ground, and *bates*, one who walks or haunts, as in *Geobates* (now *Geositta*) *peruviana*, the Coastal Miner

Geococcyx *jee-o-KOKS-siks*
Greek, *geo*, ground, and Latin, *coccyx*, from Greek *kokkyx*, cuckoo, from the bird's call, so called by the ancient Greek physician Galen because the human tailbone supposedly resembles a cuckoo's beak, as in *Geococcyx californianus*, the Greater Roadrunner

Geocolaptes *jee-o-ko-LAP-teez*
Greek, *geo*, ground and *colapt-*, chisel, peck, as in *Geocolaptes olivaceus*, the Ground Woodpecker

Geoffroyi *JEF-froy-eye*
After Geoffroy Saint-Hilaire, a French naturalist, as in *Geoffroyus geoffroyi*, the Red-cheeked Parrot

Gelochelidon *jel-o-KEL-ih-den*
Greek, *gelao*, to laugh in joy, and *chelidon*, swallow, as in *Gelochelidon nilotica*, the Gull-billed Tern. The call accounts for the laughing analogy and the wings resemble a swallow's

Geopelia *jee-o-PEL-ee-a*
Greek, *geo*, ground and *pelia*, a dove, as in *Geopelia striata*, the Zebra Dove

Geophaps *JEE-o-faps*
Greek, *geo*, ground and *phaps*, a dove or pigeon, as in *Geophaps plumifera*, the Spinifex Pigeon

Geopsittacus *jee-op-SIT-ta-kus*
Greek *geo*, ground and *psittakos*, parrot-like, as in *Geopsittacus* (now *Pezoporus*) *occidentalis*, the Night Parrot. Nocturnal and terrestrial, it is a very rare endemic of Australia

Georgiana, -us *jor-jee-AN-a/us*
After the State of Georgia in the US, as in *Melospiza georgiana*, the Swamp Sparrow

Georgica, -us *JOR-ji-ka/us*
After South Georgia, as in *Anas georgica*, the Yellow-billed Pintail

Geositta *jee-o-SIT-ta*
Greek *geo*, ground and Old English, *sittan*, to be seated, as in *Geositta peruviana*, the Coastal Miner which inhabits barren, gravelly ground, often with no vegetation

Geospiza *jee-o-SPY-za*
Greek, *geo*, ground, and *spiz-a*, finch, as in *Geospiza conirostris*, the Large Cactus Finch (see box), one of Darwin's Finches

---

LATIN IN ACTION

The genus *Geospiza* (ground finch) and four other genera, for a total of 14 species, comprise the group of birds on the Galapagos Islands known as Darwin's Finches. For many years Darwin's Finches, the ancestor(s) of which somehow made it over 1,000 kilometers of ocean from South America to establish this population, were considered as part of the Emberizidae family, which includes those birds called buntings in the Old World and sparrows in the New World. Today they are part of Thraupidae, the tanagers and relatives. Darwin observed and collected all the finches except for the Woodpecker Finch (*Camarhynchus pallidus*), but thought that they were simply variations on a type. John Gould, a famous English ornithologist, determined that they were actually separate species.

Geothlypis jee-o-thi-LIP-is
Greek, *geo*, ground, and *thlypis*, small bird, as in *Geothlypis nelsoni*, the Hooded Yellowthroat. Compared with other New World Warblers, *Geothlypis* species inhabit low vegetation

Geotrygon jee-o-TRY-gon
Greek, *gaia*, earth, and *trygon*, cooer, as in *Geotrygon chrysia*, the Key West Quail-Dove

Geranoaetus jer-an-o-EE-tus
Greek, *geranos*, crane, and *aetus*, eagle, as in *Geranoaetus albicaudatus*, the White-tailed Hawk

Geranospiza jer-an-o-SPY-za
Greek, *geranos*, crane, and *spiza*, finch, as in *Geranospiza caerulescens*, the Crane Hawk, not exactly a finch, but its gray wings and call are crane-like

Gerygone ger-IH-gon-ee
Greek, *goryo*, sound, speech, and *gone*, offspring, born of, as in *Gerygone chloronota*, the Green-backed Gerygone

Gigantea, -us jye-GAN-tee-a/us
Gigantic, as in *Fulica gigantea*, the Giant Coot

Gigas JYE-gas
Giant, as in *Patagona gigas*, the Giant Hummingbird

Gilvus JIL-vus
Pale yellow, as in *Vireo gilvus*, the Warbling Vireo

Githagineus gith-a-JIN-ee-us
Githagineus is probably a corruption of the plant species *Agrostemma githago*, the Corn Cockle, a common European flower, as in *Bucanetes githagineus*, the Trumpeter Finch, which eats its seeds

Glacialis gla-see-AL-is
Icy, as in *Fulmarus glacialis*, the Northern Fulmar, a common bird of the subarctic areas of the North Pacific and Atlantic Oceans

Glandarius glan-DAR-ee-us
Glandis, an acorn, and *arius*, quantity of, as in *Garrulus glandarius*, the Eurasian Jay, an avid acorn eater

Glareola glar-ee-O-la
Glarea, gravel, as in *Glareola pratincola*, the Collared Pratincole, which nests in a depression in the soil or gravel; common name from *prat-*, meadow, and *col-*, dwell

Glaucescens GLAW-ses-senz
Graying, as in *Larus glaucescens*, the Glaucous-winged Gull

Glaucidium, -us glaw-SID-ee-um/us
Glaucus, gray, bluish, and *dium*, open sky, as in *Glaucidium passerinum*, the Eurasian Pygmy Owl

Glaucoides glaw-KOY-deez
Glaucus, gray, bluish, and *oides*, resembling, as in *Larus glaucoides*, the Iceland Gull

Glaucus GLAW-kus
Glaucus, gray, bluish, as in *Anodorhynchus glaucus*, the Glaucous Macaw

Glossopsitta glos-sop-SIT-ta
Greek, *glosso*, tongue, and *psitta*, parrot, as in *Glossopsitta concinna*, the Musk Lorikeet

Gnoma NOM-a
Greek, gnome or dwarf, as in *Glaucidium gnoma*, the Mountain or Northern Pygmy Owl

Gnorimopsar no-ri-MOP-sar
Greek, *gnorious*, a mark, judgement, and *psar*, starling, as in *Gnorimopsar chopi*, the Chopi Blackbird, which resembles a starling

Godeffroyi god-ef-FROY-eye
After Johann Cesar Godeffroy, German zoologist, as in *Todiramphus godeffroyi*, the Marquesan Kingfisher

Godlewskii god-LOO-skee-eye
After Wiktor Godlewski, a Polish zoologist, as in *Emberiza godlewskii*, Godlewski's Bunting

Emberiza godlewskii, Godlewski's Bunting

The Color of Birds

Birds are among the most colorful of animals, their colored feathers evolving mainly as an adaptation for reproduction. Males of species such as hummingbirds, sunbirds, and tanagers attract mates with their bright plumage, and Red-winged Blackbirds, *Agelaius phoeniceus*, establish and defend territories with their blazing epaulets. And of course, in the thickness of tropical forests, the range of spectacular colors lets all of the birds know who's who. Other birds, for protection, have evolved disruptive coloration, patterns that break up their outline, such as banded plovers, and birds like nighthawks and bitterns have evolved camouflage.

Feather colors are formed by either or both pigment and structure. One pigment, melanin, produces colors from black to dull yellow; carotenoids are responsible for yellow to yellow-orange

Botaurus lentiginosus,
American Bittern

Agelaius phoeniceus,
Red-winged Blackbird

colors; and porphyrins produce bright colors in several shades of pink, red, yellow, and green. Structural colors are produced by the refraction of light through the cells of the feather. If you find a Blue Jay, *Cyanocitta cristata*, or bluebird feather and hold it in your hand, it appears blue because the incoming light is refracted as it is reflected. But if you hold the feather up to the light, the light is transmitted through the feather and it will appear brown due to the melanin granules. Iridescent colors of hummingbirds, sunbirds, and others are produced in a similar way and the angle at which the birds are viewed causes the colors to vary. Green colors are often produced by a yellow pigment deposited on top of structural blue.

Beginner birdwatchers often consider color to be the best clue to identification, being misled by the common names of birds. One would tend to look for the orange of the Orange-crowned Warbler, *Leiothlypis celata*, or an all blue Eurasian Blue Tit, *Cyanistes caeruleus*, when the orange crown is not at all obvious and the Blue Tit is not all blue.

Color perception also varies with different lighting conditions, so patterns, silhouette, behavior, and habitat are often better clues than color. Seeing color is a bonus.

But because colors are so important and so obvious a feature of birds, many of their scientific names reflect their color or color patterns. The all-white White Tern is *Gygis alba*, *Alcippe brunnea* is the mostly brown Dusky Fulvetta, and *Lonchura melanea*, the mostly black Buff-bellied Mannikin. The Blue-black Kingfisher is aptly named *Todiramphus nigrocyaneus*. Or the name may reflect the color of only a particular part as in the Little Tern, *Sternula albifrons*, with a white forehead; *Oriolus chlorocephalus*, the Green-headed Oriole; and the Cobalt-winged Parakeet, *Brotogeris cyanoptera*. There are many names that refer to color and use the color prefix like *alba-*, white, and are used repeatedly for different body parts. Hence we have *albicapilla* (white-haired), *albicauda* (white-tailed), *albiceps* (white-headed), *albicilla* (white-tailed), *albicollis* (white-collared), *albifrons* (white-forehead), etc. and *xantho*, yellow, as in *xanthogastra* (yellow belly), *xanthocollis* (yellow collar), *xanthophrys* (yellow eyebrow), etc.

The color descriptions are primarily based upon the plumage of the mature male of the species, but we often find mismatches between the descriptive scientific and common names. The Crescent Honeyeater's scientific name, *Phylidonyris pyrrhopterus*, means red or flame-colored wings when the bird's are actually bright yellow. The Myrtle (once Yellow-rumped) Warbler's specific epithet of (*Setophaga*) *coronata* refers to its crown, not its rump. The Black-billed Cuckoo's scientific name, *Coccyzus erythropthalmus*, refers to its red eye and the White-shouldered Antbird's name, *Myrmeciza melanoceps*, means black-headed.

Oriolus chlorocephalus, Green-headed Oriole

Yellow and brown pigments in the cells of feather barbules make the different shades and iridescence levels of green colors via reflection and refraction on and through the cells.

We describe and name many birds by their colors, but birds, having better vision than us, can see not only the visible spectrum of colors but also UV light. Over 90 percent of birds examined reflect UV from their feathers and probably give birds a much different view of each other than we have. Male Blue Tits raise a UV reflective crown patch during courtship and the Blue Grosbeaks, *Passerina caerulea*, with the most UV reflection in their blue feathers are the most successful breeders. The black bibs of male House Sparrows, *Passer domesticus*, indicate their level of dominance and the amount of spotting on a female Western Barn Owl's, *Tyto alba*, breast indicates her parasite load to a potential mate.

Todiramphus nigrocyaneus, Blue-black Kingfisher

Goeldii *GELD-ee-eye*
After Emil Goeldi, Swiss zoologist, as in *Myrmeciza goeldii*, Goeldi's Antbird

Goeringi *GE-ring-eye*
After Anton Goering, German naturalist and painter, as in *Brachygalba goeringi*, the Pale-headed Jacamar

Goethalsia *ge-TAL-see-a*
After George Goethals, US Army officer and chief engineer of the Panama Canal, as in *Goethalsia bella*, Pirre Hummingbird

Goffiniana *gof-fin-ee-AN-a*
After Andreas Goffin, Dutch naval officer, as in *Cacatua goffiniana*, Tanimbar Corella

Goldiei *GOLD-ee-eye*
After Andrew Goldie, Scottish explorer, as in *Psitteuteles goldiei*, Goldie's Lorikeet

Goldmania, -mani *gold-MAN-ee-a/GOLD-man-eye*
After Edward Goldman, American naturalist and mammalogist, as in *Goldmania violiceps*, Violet-capped Hummingbird

Goliath *go-LYE-ath*
Giant, Goliath, the Philistine warrior, as in *Ardea goliath*, the Goliath Heron

Goodfellowi *GOOD-fel-lo-eye*
After Walter Goodfellow, British ornithologist and explorer, as in *Regulus goodfellowi*, the Flamecrest

Goodsoni *GOOD-son-eye*
After Arthur Goodson, British ornithologist, as in *Columba* (now *Patagioenas*) *goodsoni*, the Dusky Pigeon

Goudotii *goo-DOT-ee-eye*
After Justin-Marie Goudot, French zoologist, as in *Chamaepetes goudotii*, the Sickle-winged Guan

Gouldiae, -i *GOULD-ee-ee/eye*
After John Gould, famous British ornithologist, as in *Erythrura gouldiae*, the Gouldian Finch. John Gould had 24 birds named after him, more than anyone else

Graciae *GRAY-see-ee*
After Grace Coues, sister of Elliot Coues, who first discovered *Setophaga graciae*, Grace's Warbler

Gracilirostris *gra-sil-ee-ROSS-tris*
Gracilis, slender, and *rostris*, bill, as in *Calamonastides gracilirostris*, the Papyrus Yellow Warbler

Gracilis *gra-SIL-is*
Slender, as in *Meliphaga gracilis*, the Graceful Honeyeater

Gracula, -us, -ina *gra-KOOL-a/us/gra-kool-EE-na*
Graculus, a jackdaw, as in *Gracula religiosa*, the Common Hill Myna

Gracupica *gra-koo-PIKE-a*
Graculus, a jackdaw, and *pica*, a magpie, as in *Gracupica contra*, the Pied Myna

Graduacauda *gra-doo-a-CAW-da*
Gradus, slope, walk, and *cauda*, tail, as in *Icterus graduacauda*, Audubon's Oriole, and may refer to the tapering of the tail feathers

Graeca *GREE-ka*
Graecus, Greek, as in *Alectoris graeca*, the Rock Partridge, whose home range includes Greece

Grallaria, -us *gral-LAR-ee-a/us*
Grallae, stilts and *aria*, air, as in *Fregetta grallaria*, the White-bellied Storm Petrel, from its habit of "walking" on the surface of the sea

Ardea goliath, Goliath Heron

Grallaricula *gral-lar-ih-KOOL-a*
Grallae, stilts, and *cula*, diminutive, as in *Grallaricula flavirostris*, the Ochre-breasted Antpitta, referring to its short tail making its legs appear disproportionately long

Grallina *gral-LEEN-a*
Grallae, stilts, as in *Grallina cyanoleuca*, the Magpie-lark with its longish legs

Gramineus *grah-MIN-ee-us*
Of grass, grassy, as in *Megalurus gramineus*, the Little Grassbird

Graminicola *grah-min-ih-KOL-a*
Gramineus, of grass, grassy, and *cola*, dweller, as in *Graminicola bengalensis*, the Indian Grassbird

Grammacus *GRAM-ma-kus*
Lined, striped, as in *Chondestes grammacus*, the Lark Sparrow

Grammiceps *GRAM-mi-seps*
Gramma, lines, and *ceps*, head, as in *Seicercus grammiceps*, the Sunda Warbler, with dark lines on its rufous head

Granadensis *gra-na-DEN-sis*
After New Granada, now part of present-day Colombia, as in *Picumnus granadensis*, the Grayish Piculet, a small woodpecker

Granatellus *gra-na-TEL-lus*
Granatus, garnet, as in *Granatellus venustus*, the Red-breasted Chat, with a bright-red chest

Granatina *gra-na-TEEN-a*
Granatus, garnet, as in *Erythropitta granatina*, the Garnet Pitta

Grandala *gran-DAL-a*
Grand, large, great, and *ala*, wing, as in *Grandala coelicolor*, the Grandala, with strikingly grand, as in spectacular, blue wings

Grandis *GRAN-dis*
Grand, large, great, as in *Ploceus grandis*, the Giant Weaver

Graueri, -ia *GRAU-er-eye/grau-ER-ee-a*
After Rudolph Grauer, Austrian explorer who collected in the Belgian Congo, as in *Bradypterus graueri*, Grauer's Swamp Warbler

Gravis *GRA-vis*
Heavy, important, as in *Puffinus gravis*, the Great Shearwater

Grandala coelicolor, Grandala

Grayi *GRAY-eye*
After George Gray, British ornithologist, as in *Turdus grayi*, Clay-colored Thrush; also after John Gray, older brother of George Gray, British ornithologist and entomologist, as in *Ammomanopsis grayi*, Gray's Lark

Graysoni *GRAY-son-eye*
After Andrew Jackson Grayson, American ornithologist and artist, as in *Mimus graysoni*, the Socorro Mockingbird

Grimwoodi *GRIM-wood-eye*
After Ian Grimwood, Chief Game Warden of Kenya, as in *Macronyx grimwoodi*, Grimwood's Longclaw

Grisea *GRIS-ee-a*
Griceus, gray, as in *Formicivora grisea*, the Southern White-fringed Antwren

Grisegena *grins-e-JEN-a*
Griceus, gray, and *gena*, chin, cheek, as in *Podiceps grisegena*, the Red-necked Grebe, with grayish-white cheeks

Griseicapilla, -us *gris-ee-eye-ka-PIL-la/us*
Griceus, gray, and *capilla*, hair on the head, as in *Sittasomus griseicapillus*, the Olivaceous Woodcreeper

Griseiceps *gris-ee-EYE-seps*
Griceus, gray, and *ceps*, head, as in *Accipiter griseiceps*, the Sulawesi Goshawk

Griseicollis *gris-ee-eye-KOL-lis*
Griceus, gray, and *collis*, the neck, as in *Scytalopus griseicollis*, the Pale-bellied Tapaculo

Griseigula, -gularis *gris-ee-eye-GOO-la/ gris-ee-eye-goo-LAR-is*
Griceus, gray, and *gula*, the throat, as in *Timeliopsis griseigula*, the Tawny Straightbill

Griseipectus *gris-ee-eye-PEK-tus*
Griceus, gray, and *pectis*, the breast, as in *Pyrrhura griseipectus*, the Gray-breasted Parakeet

Griseiventris *gris-ee-eye-VEN-tris*
Griceus, gray, and *ventris*, the underside, belly, as in *Melaniparus griseiventris*, the Miombo Tit

Griseocephalus *gris-ee-o-se-FAL-us*
Griceus, gray, and Latin, *cephala*, head, as in *Dendropicos griseocephalus*, the Olive Woodpecker

Griseoceps *gris-ee-O-seps*
Griceus, gray, and *ceps*, head, as in *Microeca griseoceps*, the Yellow-legged Flyrobin

Griseogularis *gris-ee-o-goo-LAR-is*
Griceus, gray, and *gularis*, throated, as in *Ammoperdix griseogularis*, the See-see Partridge

Griseus *GRIS-ee-us*
Gray, as in *Nyctibius griseus*, the Common Potoo, common name after its wailing call

Grossus *GRO-sus*
Grossus, thick, as in *Saltator grossus*, the Slate-colored Grosbeak, with a thick beak

Grus *GRUSS*
Crane, such as *Grus americana*, the Whooping Crane

Grylle *GRIL-lee*
Scottish name for *Cepphus grylle*, the Black Guillemot

Gryphus *GRIF-us*
Greek, *gryp-*, hook-nosed, as in *Vultur gryphus*, the Andean Condor

Guadalcanaria *gwa-dal-kan-AR-ee-a*
After Guadalcanal Island, in the Solomon Islands, as in *Guadalcanaria inexpectata*, the Guadalcanal Honeyeater

Guarauna *gwa-RAWN-a*
The Brazilian Indian name for this bird, *Aramus guarauna*, the Limpkin, the common name coming from the bird's limping gait

Gubernetes *goo-ber-NEET-eez*
A rudder, governor, as in *Gubernetes yetapa*, the Streamer-tailed Tyrant

Gujanensis *goo-ja-NEN-sis*
After French Guinea, as in *Odontophorus gujanensis*, the Marbled Wood Quail

Gularis *goo-LAR-is*
Gula, throat, gullet, as in *Egretta gularis*, the Western Reef Heron or Egret; *Gularis* probably refers to the large throat of these birds, and there are two dozen with this specific epithet

Gurneyi *GER-nee-eye*
After John Gurney, British banker and amateur ornithologist, as in *Aquila gurneyi*, Gurney's Eagle

Grus americana, Whooping Crane

LATIN IN ACTION

Ornithologists know little about the Bornean Bristlehead, *Pityriasis gymnocephala*. Presently considered the only member of the Pityriaseidae family and genus *Pityriasis*, in the past it was placed in other families, including Corvidae, the jay and crow family. It is a rainforest inhabitant but due to the destruction of forests by logging and the illegal black market for these birds as pets, it is considered near threatened. This iconic bird is the most sought after by birdwatchers in Borneo.

Pityriasis gymnocephala, Bornean Bristlehead

Gymnocichla *jim-no-SICK-la*
Greek, *gymno*, naked, bare, and *cichla*, thrush, as in *Gymnocichla nudiceps*, the Bare-crowned Antbird

Gymnoderus *jim-no-DER-us*
Greek, *gymno*, naked, bare, and *der-*, neck, hide, as in *Gymnoderus foetidus*, the Bare-necked Fruitcrow

Gymnoglaux *JIM-no-glawks*
Greek, *gymno*, naked, bare, and *glaux*, owl, as in *Gymnoglaux* (now *Margarobyas*) *lawrencii*, the Bare-legged Owl

Gymnogyps *JIM-no-jips*
Greek, *gymno*, naked, bare, and *gyps*, vulture, as in *Gymnogyps californianus*, the California Condor; Condor derives from American Spanish, *cuntur*, the native name for the bird

Gymnorhinus, -a *jim-no-RYE-nus/na*
Greek, *gymno*, naked, bare, and *rhinos*, nose, as in *Gymnorhinus cyanocephalus*, the Pinyon Jay, whose bill is featherless at the base

Gypaetus *ji-PEE-tus*
Greek, *gymno*, naked, bare, and *aetus*, eagle, as in *Gypaetus barbatus*, the Bearded Vulture

Gyps *JIPS*
Greek, *gyps*, vulture, as in *Gyps fulvus*, the Griffon Vulture

Guttata, -us *gut-TAT-a/us*
Gutta, drop, spot, speck, as in *Ortalis guttata*, the Speckled Chachalaca, whose common name comes from its loud calls

Guttaticollis *gut-ta-ti-KOL-lis*
Gutta, drop, spot, speck, and *collis*, neck, as in *Paradoxornis guttaticollis*, the Spot-breasted Parrotbill

Gutturalis *gut-ter-AL-is*
Guttur, the throat, as in *Anthus gutturalis*, the Alpine Pipit, with a streaked throat

Guy *GEE*
After J. Guy, French naturalist, as in *Phaethornis guy*, the Green Hermit

Gygis *JI-jis*
Guges, a water bird, as in *Gygis alba*, the White Tern

Gymnocephala, -us *jim-no-se-FAL-a/us*
Greek, *gymno*, naked, bare, and Latin, *cephala*, head, as in *Pityriasis gymnocephala*, the Bornean Bristlehead (see box)

Gymnogyps californianus, California Condor

PHOEBE SNETSINGER

(1921–1999)

Phoebe Snetsinger was born Phoebe Burnett in 1921 in Lake Zurich, Illinois. Her father, Leo Burnett, was the advertising executive who made famous the Jolly Green Giant, the Marlboro Man, Toucan Sam, Charlie the Tuna, Morris the Cat, the Pillsbury Doughboy, and Tony the Tiger. His successes and the resulting financial rewards eventually enabled Phoebe to travel the world in search of birds. Only eight bird-watchers in history have ever seen more than 8,000 of the approximately 10,000 species of birds found on our planet. Phoebe Snetsinger, of Missouri, was one of the eight.

When she started keeping a list, there were 8,500 officially named species, compared with about 10,000 now. Her list of more than 2,000 bird genera far surpassed anyone else's, and she was especially interested in monotypic genera, those genera that contain only one species of bird. She also kept notes on subspecies and geographic races that have since been elevated to the species level. So her life list of 8,400 species continues to grow even after her death in 1999.

She married her husband David Snetsinger, a scientist and administrator, whom she had known since the age of eleven. The marriage proved unfulfilling for Snetsinger, so she and her husband drifted apart but didn't divorce. She wrote dark, despairing poems, describing her marriage as "a stodgy, graceless, larval time."

Calicalicus rufocarpalis,
Red-shouldered Vanga

The rare Red-shouldered Vanga is endemic to south-western Madagascar and may be best known as the last bird to be sighted by Phoebe Snetsinger.

When Snetsinger was 34, a friend introduced her to bird-watching, and the sight of a Blackburnian Warbler, *Setophaga fusca*, changed her life. With her photographic memory and a fierce will to learn, she proved an excellent birder. Birding went from a hobby to a passion for Snetsinger in 1981 when a doctor told her she had terminal melanoma cancer and a short time to live. Rejecting therapy, she took off to Alaska on a scheduled trip, her first long-distance journey simply to see birds. She was 49.

Snetsinger liked to say her love of birds "began with a death sentence," and her relentless energy reflected that level of urgency as her cancer went into the first of several remissions. "Birding has meant a variety of things to many different people," Snetsinger once wrote in an article for a nature club, "but for me it has been intricately intertwined with survival." After her diagnosis she spent more time in the wilds of the world—jungles, swamps, deserts—than she did at home. She was most comfortable with her binoculars, floppy hat, and notebook.

Many of her birding tours cost more than $5,000, and she maintained this travel schedule for 18 years after her diagnosis! There were setbacks, as the

"You'd go that far to see one bird?"

Phoebe Snetsinger

melanoma recurred every five years or so, but always went into remission again. She died in an auto accident on a birding expedition to Madagascar, shortly after viewing an exceptionally rare Helmet Vanga, *Euryceros prevostii*, or Red-shouldered Vanga, *Calicalicus rufocarpalis*, depending on the story one reads. She was 68.

Well, with a name like Phoebe, she almost had to be a birder. There are only 900 species in North America, so she had to have the time and money to travel extensively on other continents to reach her 8,400 species. A few other people have gone as far in pursuit of birds, but only about 250 of them have ever hit the 5,000 mark, perhaps 100 people have seen 6,000 and only 12 or so have seen more than 7,000.

At the time of writing Tom Gullick, a British resident of Spain, is the only person ever to see 9,000 bird species; the 9,000th was the endemic Wallace's Fruit-Dove, *Ptilinopus wallacii*, on Yamdena in the Tanimbar Islands, Indonesia. He ended that trip with a total of 9,047 species. Gullick has been the top world lister since 2008 and holds the record for the most species seen in South America (2,939) and Africa (2,081).

But Phoebe Snetsinger has to be given credit for her passion and perseverance, qualities that made her the icon of diehard bird listing. At the time of her death she was 2,000 birds ahead of her nearest rival. Some of her adventures and misadventures, which included recurrences of cancer, a gang rape in New Guinea, a shipwreck, earthquakes, and political problems, are detailed in her book *Birding on Borrowed Time*, published in 2003.

Setophaga fusca,
Blackburnian Warbler

The Blackburnian Warbler is not easy to spot as it prefers to forage in the treetops, where it searches the branch tips for insects and larvae.

Aramides axillaris,
Rufous-necked Wood Rail

Snetsinger passed the 8,000 mark in September 1995 when she spotted her first Rufous-necked Wood Rail.

HAASTII

Haastii *HAAST-ee-eye*
After Johann Franz "Julius" von Haast, a German geologist who worked in New Zealand, as in *Apteryx haastii*, the Great Spotted Kiwi

Habia *HA-bee-a*
From an indigenous language of South America (Guaraní) as in *Habia rubica*, the Red-crowned Ant Tanager

Habroptila *ha-brop-TIL-a*
Greek, *habro*, dainty, delicate, and *ptila*, feather, as in *Habroptila wallacii*, the Invisible Rail

Haemacephala *hee-ma-se-FAL-a*
Greek, *haima*, blood, and Latin, *cephala*, head, as in *Megalaima haemacephala*, the Coppersmith Barbet

Haemastica *hee-MASS-tik-a*
Greek, *haima*, blood, as in *Limosa haemastica*, the Hudsonian Godwit, with chestnut-red underparts

Haematoderus *hee-ma-to-DER-us*
Greek, *haimo*, blood, and *dera*, neck, throat, as in *Haematoderus militaris*, the Crimson Fruitcrow

Haematogaster *hee-ma-to-GAS-ter*
Greek, *haimo*, blood, and *gaster*, stomach, as in *Campephilus haematogaster*, the Crimson-bellied Woodpecker

Haematonota, -us *hee-ma-toe-NO-ta/tus*
Greek, *haimo*, blood, and *noto*, back, as in *Epinecrophylla haematonota*, the Stipple-throated Antwren

Haematopus *hee-ma-TO-pus*
Greek, *haimo*, blood, and *pous*, foot, as in *Haematopus ater*, the Blackish Oystercatcher, although the bill is blood red, not the feet

Haematortyx *hee-ma-TOR-tiks*
Greek, *haimo*, blood, and *ortux*, quail, as in *Haematortyx sanguiniceps*, the Crimson-headed Partridge

Haematospiza *hee-ma-to-SPY-za*
Greek, *haimo*, blood, and *spiza*, finch, as in *Haematospiza* (now *Carpodacus*) *sipahi*, the Scarlet Finch

Hainanus *hye-NAN-us*
After Hainan, China, as in *Cyornis hainanus*, the Hainan Blue Flycatcher

Halcyon *HAL-see-on*
Greek, kingfisher, as in *Halcyon senegalensis*, the Woodland Kingfisher

Haliaeetus *hal-ee-a-EE-tus*
Greek, *hals*, the sea, and *aetus*, eagle, as in *Haliaeetus leucogaster*, the White-bellied Sea Eagle

Haliaetus *ha-lee-EE-tus*
Sea eagle, osprey, as in *Pandion haliaetus*, the Western Osprey

Haliastur *ha-lee-AST-ur*
Greek *hals*, the sea, and *-astur*, a hawk, as in *Haliastur indus*, the Brahminy Kite, which often feeds along the coast

Halli *HALL-eye*
After Robert Hall, Australian ornithologist, as in *Macronectes halli*, Northern Giant Petrel

Habroptila wallacii,
Invisible Rail

Halcyon

There are about 90 species of kingfishers spread over 17 genera. The *Halcyon* (*HAL-see-on*) genus contains 11 of the 60 or so bird species known as tree or wood kingfishers and are primarily Old World in distribution. *Halcyon* comes from Alcyone of Greek mythology, daughter of Aeolus, the ruler of the winds. She married Ceyx, who died in a shipwreck. Alcyone was so upset she drowned herself in the sea, after which the gods turned both of them into kingfishers. When Alcyone nested, Aeolus calmed the winds for a week. These seven days became known as the "halcyon days."

Kingfishers are so called because they are supposedly the "king of the fishers," but tree kingfishers will also take small reptiles, amphibians, crabs, and even small birds and mammals. The Ruddy Kingfisher (*H. coromanda*) is known to feed on land snails that it crushes with an "anvil rock." Kingfishers typically beat larger prey on a branch to disable and soften it before swallowing.

Tree kingfishers will nest in a tree cavity made by woodpeckers or dig out rotting wood to make a hole. Some will nest in termite nests and others excavate tunnels in riverbanks. Like all the other members of its avian order Coraciiformes, which includes bee eaters, rollers, and hornbills, their feet exhibit "syndactly" (fused toes); their third and fourth toes are joined at the base to help them to dig nest holes. They are monogamous and territorial; along river banks these territories are likely to be long and narrow, but for the forest-nesting species they are oval or circular. Kingfishers lay four to seven eggs in a nest cavity 20 to 40 inches (50 to 100 centimeters) long. In times when food is scarce, egg-laying may take place every other day but incubation begins immediately, so the young are at different ages and sizes as they hatch. The older hatchlings are more successful in begging for food and therefore have a better chance of survival than the younger ones. This strategy of asynchronous hatching is also employed by birds of prey and other birds to assure that at least one or two young make it to fledging.

Halcyon senegaloides, Mangrove Kingfisher

Halcyon coromanda, Ruddy Kingfisher

Haliaeetus

From the Greek for sea eagle or osprey, the genus *Haliaeetus* (*ha-lee-EE-tus*) contains eight living species and is one of the oldest groups of birds, commonly known as sea eagles. Most have white tails and a few have white heads. Perhaps the most well-known, weighing in at 13 pounds (6 kilograms), is the Bald Eagle, *H. leucocephalus*, the national bird of the US. The bird is not really bald; its name derives from the term "piebald," which refers to large patches of color, usually white.

Most sea eagles feed on fish but will take other prey and are not averse to eating carrion. In Alaska, where there are no vultures, Bald Eagles can be seen scavenging around garbage dumps. Sea eagles will also harass other birds such as gannets and gulls in an effort to make them drop their piscine prey. But they are also efficient predators. The White-bellied Sea Eagle, *H. leucogaster*, flies low over the water with its talons tucked under its chin and strikes rapidly at the water surface while flapping its wings in a strong effort to take off once the fish

Haliaeetus leucocephalus,
Bald Eagle

is seized. The African Fish Eagle, *H. vocifer*, flies from its perch in a tree to swoop down on fish and, like all sea eagles, has prickles on the underside of its toes to help hold its slippery prey. The White-tailed Eagle, *H. albicilla*, eats a variety of fish but commonly targets water birds such as terns, cormorants, loons (or divers), grebes, ducks, coots, and even skuas.

Sea eagles, mature by about five years of age, mate for years, sometimes even for life, according to evolving evidence. The pairs build huge nests that may exceed 10 feet (3 meters) in diameter and weigh 3 tons. The nests may be used year after year for many years, sometimes by successive generations.

The populations of sea eagles in both North America and Europe have suffered because they are top predators and accumulate toxins, such as pesticides and pollutants. They have also been shot and harassed by farmers, hunters, and egg-collectors over the years, accelerating the decline.

Haliaeetus vocifer,
African Fish Eagle

Halobaena *ha-lo-BEEN-a*
Greek, *hals*, the sea, and *baen*, walk, step, as in *Halobaena caerulea*, the Blue Petrel, for the petrel habit of "tip-toeing" on the ocean's surface

Halocyptena *ha-lo-sip-TEN-a*
Greek, *hals*, sea, *okus*, speedy, and *ptenos*, winged, flying, as in *Halocyptena* (now *Oceanodroma*) *microsoma*, the Least Storm Petrel

Hamirostra *ha-mee-ROSS-tra*
Hamus, hooked, and *rostris*, beak, as in *Hamirostra melanosternon*, the Black-breasted Buzzard

Hammondii *ham-MOND-ee-eye*
After William Hammond, military physician and biological collector, as in *Empidonax hammondii*, Hammond's Flycatcher

Hapalopsittaca *ha-pa-lop-SIT-ta-ka*
Greek, *hapalo*, gentle, soft, and Latin, *psittaca*, parrot, as in *Hapalopsittaca amazonina*, the Rusty-faced Parrot

Hapaloptila *ha-pa-lop-TIL-a*
Greek, *hapalo*, gentle, soft, and *ptilon*, feather, as in *Hapaloptila castanea*, the White-faced Nunbird

Haplochelidon *hap-lo-kel-EYE-don*
Greek, *hapalo*, gentle, soft, and *chelidon*, swallow, as in *Haplochelidon andecola*, the Andean Swallow

Haplochrous *hap-LO-krus*
Greek, *hapalo*, gentle, soft, and *chroa*, skin, complexion, as in *Accipiter haplochrous*, the White-bellied Goshawk, from the soft appearance of its plumage, especially the white belly

Haplonota *hap-lo-NO-ta*
Greek, *hapalo*, gentle, soft, and *notos*, back, as in *Grallaria haplonota*, the Plain-backed Antpitta

Haplophaedia *hap-lo-FEE-dee-a*
Greek, *hapalo*, gentle, soft, and *phaedros*, bright, brilliant, as in *Haplophaedia lugens*, the Hoary Puffleg

Haplospiza *hap-lo-SPY-za*
Greek, *hapalo*, gentle, soft, and *spiza*, finch, as in *Haplospiza rustica*, the Slaty Finch, from the soft appearance of its plumage

Hardwickii *hard-WIK-ee-eye*
After Thomas Hardwicke, General in the East India Company, as in *Chloropsis hardwickii*, the Orange-bellied Leafbird

Harpactes *har-PAK-teez*
Greek, *harpact*, to rob, seize, as in *Harpactes ardens*, the Philippine Trogon, which steals the nests of termites and wasps to use as its own. *Trogon* is Greek for nibbler: it gnaws at tree bark to make cavities

Harpagus *har-PAY-gus*
Greek, *harpag*, hook, as in *Harpagus bidentatus*, the Double-toothed Kite

Harpia *HAR-pee-a*
Greek, *harpi*, a sickle, bird of prey, as in *Harpia harpyja*, the Harpy Eagle, after the mythical harpy

Harpyhaliaetus *har-pee-hal-ee-EE-tus*
Greek, *harpi*, a sickle, bird of prey, and *haliaet, -e, -us*, sea eagle, osprey, as in *Harpyhaliaetus* (now *Buteogallus*) *coronatus*, the Crowned Solitary Eagle

Harpyopsis *har-pee-OP-sis*
Greek, *harpi*, a sickle, bird of prey, and *opsis*, appearance, as in *Harpyopsis novaeguineae*, the Papuan Eagle

Harterti, -tula *HART-ert-eye/hart-er-TOO-la*
After Ernst Hartert, German ornithologist, as in *Asthenes harterti*, the Black-throated Thistletail

Hartlaubi, -ii *HART-laub-eye/hart-LAUB-ee-eye*
After Karl Hartlaub, German academic and explorer, as in *Tauraco hartlaubi*, Hartlaub's Turaco

Hapalopsittaca amazonina, Rusty-faced Parrot

LATIN IN ACTION

Being the smallest bird in the world is both a distinction and a burden. The Bee Hummingbird (*Mellisuga helenae*), once called the Cuban Bee Hummingbird, has a very high metabolism because of its size, about 2 to 2½ inches (5 to 6 centimeters) long and weighing about 1.7 grams, comparable in size to a large bee. Some amateur birdwatchers have mistaken bees and moths for Bee Hummingbirds. The small size means that the large surface area of the bird, which is responsible for heat loss, and the small volume of the bird, which produces body heat, requires that the bird spend 15 percent of its daily activity eating. Its daily body temperature is 105°F (41°C) but drops to 86°F (30°C) at night to conserve energy. They could not survive without going into nightly torpor. *Mellisuga* (honey sucker) is a bit misleading because they actually ingest nectar, not honey, and not by sucking, but by sopping it up with their mop-like tongue.

Heermanni *HAIR-man-nye*
After Adolphus Heermann, an American military surgeon-naturalist, as in *Larus heermanni*, Heermann's Gull

Heinrichia, -i *hine-RICK-ee-a/eye*
After Gerd Heinrich, German zoologist, as in *Heinrichia calligyna*, the Great Shortwing

Heinrothi *HINE-rot-eye*
After Oskar Heinroth, German zoologist, as in *Puffinus heinrothi*, Heinroth's Shearwater

Heleia *hel-LAY-ee-a*
Greek, Helen, as in *Heleia muelleri*, the Spot-breasted Heleia

Helenae *HEL-en-ee*
Greek, Helen, as in *Mellisuga helenae*, the Bee Hummingbird, smallest bird in the world; Helenae probably from Helen Booth, wife of Charles Booth, British philanthropist

Heliactin *hel-ee-ACT-in*
Greek, *helios*, sun, and *actis*, a ray, beam, as in *Heliactin bilophus*, the Horned Sungem

Heliangelus *hel-ee-an-JEL-us*
Greek, *helios*, sun, and *angelus*, a messenger or angel, as in *Heliangelus mavors*, the Orange-throated Sunangel

Harwoodi *HAR-wood-eye*
After Leonard Harwood, English naturalist and taxidermist, as in *Pternistis harwoodi*, Harwood's Francolin

Hasitata *has-ih-TA-ta*
Hesitate, as in *Pterodroma hasitata*, the Black-capped Petrel, alluding to the first observer's uncertainty about naming the bird

Hauxwelli *HAWKS-wel-lye*
After J. Hauxwell, English bird collector, as in *Turdus hauxwelli*, Hauxwell's Thrush

Hawaiiensis *ha-wy-ee-EN-sis*
After Hawaii, as in *Corvus hawaiiensis*, the now extinct Hawaiian Crow, which is extinct in the wild

Hedydipna *hed-ee-DIP-na*
Greek, *hedy*, sweet, *dipna*, meal, as in *Hedydipna collaris*, the Collared Sunbird, which feeds on nectar

Pterodroma hasitata, Black-capped Petrel

Heliomaster longirostris,
Long-billed Starthroat

Helianthea *hel-ee-AN-thee-a*
Greek, *helios*, sun, and *anthea*, flower, as in *Coeligena helianthea*, the Blue-throated Starfrontlet

Helias *HEL-ee-as*
Greek, *helios*, sun, as in *Eurypyga helias*, the Sunbittern. The pattern on its outspread wings resembles a rising sun

Heliobates *hel-ee-o-BA-teez*
Greek, *helios*, sun, and *bates*, one that walks or hunts, as in *Camarhynchus heliobates*, the Mangrove Finch; inhabits the Galapagos, a very sunny place

Heliobletus *hel-ee-o-BLE-tus*
Greek, *helios*, sun, and *bletos*, affected, hurt, as in *Heliobletus contaminatus*, the Sharp-billed Treehunter; the sun beats down heavily on this bird

Heliodoxa *hel-ee-o-DOK-sa*
Greek, *helios*, sun, and *doxa*, glory, as in *Heliodoxa gularis*, the Pink-throated Brilliant

Heliomaster *hel-ee-o-MASS-ter*
Greek, *helios*, sun, and *master*, to shine, as in *Heliomaster longirostris*, the Long-billed Starthroat

Heliopais *hel-ee-o-PYE-is*
Greek, *helios*, sun, and *paid*, child, as in *Heliopais personatus*, the Masked Finfoot; the reference to child may have to do with the bird's ability to fly with young tucked into wing pouches

Heliornis *hel-ee-OR-nis*
Greek, *helios*, sun, and *ornis*, bird, as in *Heliornis fulica*, the Sungrebe, the name coming from the markings on the underside of the wings that resemble suns

Heliothryx *hel-ee-O-thriks*
Greek, *helios*, sun, and *thrix*, hair, as in *Heliothryx auritus*, the Black-eared Fairy; *thrix* probably refers to the delicate feathering

Hellmayri *HEL-mare-eye*
After Charles Hellmayr, a German zoologist, as in *Anthus hellmayri*, Hellmayr's Pipit

Helmitheros *hel-MIH-ther-os*
Greek, *helmins*, worm, and *theros*, hunt, as in *Helmitheros vermivorum*, the Worm-eating Warbler

Heloisa *hel-o-EE-sa*
Heloise, French name, as in *Atthis heloisa*, the Bumblebee Hummingbird. Who Heloise was is unclear; here, probably the nun of Abelard and Heloise fame

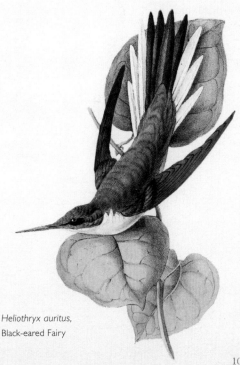

Heliothryx auritus,
Black-eared Fairy

Hemicircus concretus,
Gray-and-buff Woodpecker

Hemicircus *heh-mee-SIR-kus*
Greek, *hemi-*, half, and *circus*, ring, circle, as in *Hemicircus concretus*, the Gray-and-buff Woodpecker, with scalloped feathering on its wings

Hemignathus *heh-mig-NATH-us*
Greek, *hemi-*, half, and *gnathus*, jaw, as in *Hemignathus* (now *Chlorodrepanis*) *virens*, the Hawaii Amakihi; the lower jaw is about half as long as the upper jaw

Hemileucurus *heh-mi-loy-KOO-rus*
Greek, *hemi-*, half, *leucos*, white, and *oura*, tail, as in *Campylopterus hemileucurus*, the Violet Sabrewing

Hemileucus *heh-mi-LOY-kus*
Greek, *hemi-*, half, and *leuc-*, white, as in *Lampornis hemileucus*, the White-bellied Mountaingem

Hemimacronyx *heh-mi-ma-KRON-iks*
Greek, *hemi-*, half, *makros*, large, long, and *onux*, claw, as in *Hemimacronyx* (now *Anthus*) *chloris*, the Yellow-breasted Pipit; *hemi-* refers to its (half/partial) relationship to the closely related but debated genus *Macronyx*, rather than the claw

Hemiphaga *heh-mee-FAY-ga*
Greek, *hemi-*, half, and *phagein*, eat, as in *Hemiphaga chathamensis*, the Chatham Pigeon. Shortened from *Hemicarpophaga*, eater of half seeds, a description of its feeding habit (disperser of undigested seeds)

Hemiprocne *heh-mee-PROK-nee*
Greek, *hemi-*, half, and Latin, *progne*, swallow, as in *Hemiprocne coronata*, the Crested Treeswift; it resembles a swallow but is in a different family

Hemipus *HEM-ih-pus*
Greek, *hemi-*, half, and *pous*, foot, as in *Hemipus picatus*, the Bar-winged Flycatcher-shrike, with smaller legs and feet than birds of a similar size in the same family

Hemispingus *hem-ee-SPIN-gus*
Greek, *hemi-*, half, and *spingus*, sparrow, as in *Hemispingus reyi*, the Gray-capped Hemispingus, a warbler-like tanager

Hemitesia *hem-ee-TESS-ee-a*
Greek, *hemi-*, half, and *tesia*, a genus of warbler, as in *Hemitesia* (now *Urosphena*) *neumanni*, Neumann's Warbler

Hemithraupis *hem-ee-THRAW-pis*
Greek, *hemi-*, half, and *thraupis*, a small bird, as in *Hemithraupis ruficapilla*, the Rufous-headed Tanager

Hemixantha *hem-iks-AN-tha*
Greek, *hemi-*, half, and *xanth*, yellow, as in *Microeca hemixantha*, the Golden-bellied Flyrobin

Hendersoni *HEN-der-son-eye*
After George Henderson, English Army officer and traveler, as in *Podoces hendersoni*, Henderson's Ground Jay

Henslowii *henz-LOW-ee-eye*
After John Henslow, English botanist, as in *Ammodramus henslowii*, Henslow's Sparrow

Herberti *HER-bert-eye*
After E. G. Herbert, English collector and naturalist, as in *Stachyris herberti*, the Sooty Babbler

Herbicola *her-bi-KO-la*
Herbi, grass, and *cola*, dwell, as in *Emberizoides herbicola*, the Wedge-tailed Grass Finch

Herodias *heh-ROD-ee-us*
Greek, heron, as in *Ardea herodias*, the Great Blue Heron

Herpetotheres *her-pe-to-THER-eez*
Greek, *herpeto*, reptile, and *thero*, hunt, as in *Herpetotheres cachinnans*, the Laughing Falcon

Herpsilochmus *herp-si-LOK-mus*
Greek, *herpso*, creep, creeping, and *lochmus*, thicket *Herpsilochmus gentryi*, the Ancient Antwren

Ardea herodias,
Great Blue Heron

Phillip Clancey
(1917–2001)

Born in 1917 in Glasgow, Scotland, Phillip Clancey was educated there and developed his artistic skills at the Glasgow School of Art. He showed an early interest in birds and joined the British Ornithologists' Union at the age of 20. Over the next 16 years he published a variety of papers on the systematics of birds, especially those of Scotland. Thirty-three of his holotypes and 5,500 Western Palearctic bird skins he collected are now housed at the National Museum of Scotland.

Clancey served with the allied forces in the British Army in Sicily and Italy in World War II, and was deafened in one ear by an artillery explosion. Despite the hardships of war he pursued his avocation during the conflict and collected a race of the Woodchat Shrike (*Lanius senator*) in Sicily.

In 1948 and 1949 he accompanied Col. Richard Meinertzhegen as a field assistant on an ornithological expedition to Yemen, Aden, Somalia, Ethiopia, Kenya, and South Africa. At one point Meinertzhegen and Clancey had a heated argument about bustards in Namibia that became so violent they drew guns on each other. The bird skinner intervened to defuse the situation. Another time Meinertzhegen abandoned Clancey when he was very ill. Meinertzhegen later published the findings of this expedition in *Birds of Arabia* without ever mentioning Clancey's considerable contributions to the research.

Clancey immigrated to South Africa in 1950 and was hired as the curator of the Natal Museum in Pietermartizburg, even though he had no formal education beyond secondary school. In 1952 he became the director of the Durban Museum and Art Gallery, a position he held until his retirement in 1982. He also served as president of the Southern African Museum Association, president of the Southern Africa Ornithological Society, and was long-standing president of the Natal Bird Club. The American Ornithologists' Union honored him by naming him corresponding fellow.

Woodchat Shrike,
Lanius senator

Latin, *Lanius*, means butcher, and *senator* refers to the senatorial robe-like pattern of the male's back.

During his tenure as director of the Durban Museum and Art Gallery, Clancey participated in, initiated, and led 32 ornithological expeditions to various parts of Southern Africa. He compiled a large number of new distribution records and collected many specimens for the museum. His expeditions to Mozambique were especially important because he succeeded in bringing back the largest number of specimens ever collected from this country. He donated his collection of nearly 32,000 bird skins, considered to be the finest in Africa, to the museum. He prepared many of these skins himself and was noted for his expertise in this area. Unfortunately, he was not the most ethical of collectors; he was criticized for his disregard for restrictions stated on collecting permits. At one point, he was arrested for collecting without a permit and his shotgun was confiscated. He later bought that same shotgun back at an auction.

Clancey wrote and published extensively, amassing over 600 publications, several of which were substantial and respected works, such as *The Birds of Natal and Zululand* (1964), *The Rare Birds of Southern Africa* (1985), and *The Birds of Southern Mozambique* (1996), which he illustrated with his own paintings. Clancey also co-authored the second volume of the *Atlas of Speciation of African Birds*, published by the British Natural History Museum in 1978.

Troglodytes troglodytes indigenus, Eurasian Wren

Named *Troglodytes troglodytes indigenus* by Clancey.

He was awarded an honorary Doctor of Science from the University of Natal, the Gill Memorial Medal of the Southern African Ornithological Society, and a fellowship from the Museum Association in London, England. He named over 200 subspecies of African birds and several avian subspecies were named after Clancey by other ornithologists. He continued as a research associate of the Durban Museum and Art Gallery until his death in 2001 at the age of 83.

A life-long bachelor, Clancey focused almost exclusively on his ornithological and museum work, but in later years he increasingly devoted himself to painting and taking on commissions. He was so adept at painting that when someone for whom he had painted a bird pointed out that the eye color was wrong, Clancey changed it with one brushstroke. His artistic talents are evident not only in the many bird paintings presented in his books, but also in the dioramas on display in the Durban Natural Science Museum. He repainted some of the dioramas up to six times until they met his standards. His portraits of birds are still in great demand by collectors.

> "Clancey was a rare combination of scientist, author, artist, and administrator."
>
> David Allan, The Auk (2003)

Hesperiphona hess-pear-ih-PHONE-a
Greek, *hesperis*, evening, and *phone*, sound, as in *Hesperiphona vespertina*, the Evening Grosbeak

Heterocercus he-ter-o-SIR-kus
Greek, *heteros*, different, and *cerco*, tail, as in *Heterocercus flavivertex*, the Yellow-crested Manakin. The tails of this Manakin genus are different from other Manakin genera

Heterolaemus he-ter-o-LEE-mus
Greek, *heteros*, different, and *laemus*, throat as in *Phyllergates heterolaemus*, the Rufous-headed Tailorbird; the white throat stands out from the rest of the plumage

Heteromyias he-ter-o-MY-ee-as
Greek, *heteros*, different, and *muia*, fly, as in *Heteromyias albispecularis*, the Ashy Robin. Probably refers to the slightly different flycatcher diet of this bird

Heteronetta he-ter-o-NET-ta
Greek, *heteros*, different, and *netta*, duck, as in *Heteronetta atricapilla*, the Black-headed Duck; an unusual duck, falling somewhere between the diving and stiff-tailed ducks

Heterophasia he-ter-o-FAZ-ee-a
Greek, *heteros*, different, and *phasia*, speech, as in *Heterophasia auricularis*, the White-eared Sibia; presumably named for its call

Heteroscelus heh-ter-os-SEL-us
Greek, *heteros*, different, and *skelos*, leg, as in *Heteroscelus brevipes*, the Gray-tailed Tattler, with different scaling on its legs; new DNA evidence puts the bird into the genus *Tringa*

Heuglinii, -i, hoy-GLIN-ee-eye/HOY-glin-eye
After Theodor von Heuglin, German engineer and ornithologist, as in *Neotis heuglinii*, Heuglin's Bustard

Hiaticula hy-at-ih-KUL-a
Hiatus, cleft, opening, and *cula*, inhabit, dwell, as in *Charadrius hiaticula*, the Common Ringed Plover

Hildebrandti HIL-de-brant-eye
After Johann Hildebrandt, German collector, as in *Pternistis hildebrandti*, Hildebrandt's Francolin

Himantopus him-an-TO-pus
Greek, *himanto*, strap, and *pous*, foot, as in *Himantopus mexicanus*, the Black-necked Stilt, with long legs

Himantornis him-an-TOR-nis
Greek, *himanto*, strap, and *ornis*, bird, as in *Himantornis haematopus*, the Nkulengu Rail

Himatione hih-ma-tee-OWN-ee
Greek, cloak, as in *Himatione sanguinea*, the Apapane, whose plumage looks like a red cloak

Hirsuta, -us her-SOOT-a/us
Hairy, rough, as in *Glaucis hirsutus*, the Rufous-breasted Hermit; immature birds of this species have hairy-looking throats

Hirundapus here-un-DAP-us
Hirund, swallow, and Greek, *pous*, foot, as in *Hirundapus celebensis*, the Purple Needletail. Swifts and swallows resemble each other and have small feet

Hirundinacea, -us, -um here-un-di-NACE-ee-a/us/um
Like a swallow, as in *Euphonia hirundinacea*, the Yellow-throated Euphonia

Hirundo here-UN-do
Swallow, as in *Hirundo rustica*, the Barn Swallow

Hispaniolensis hiss-pan-ee-o-LEN-sis
After Hispaniola, as in *Contopus hispaniolensis*, the Hispaniolan Pewee

Histrionicus hiss-tree-ON-ih-kus
Histrionic, from *histro*, actor, as in *Histrionicus histrionicus*, the Harlequin Duck, referring to its brightly patterned, clown-like feathers

Hodgsoni HOJ-son-eye
After Brian Hodgson, an official of the East India Company, as in *Phoenicurus hodgsoni*, Hodgson's Redstart

Himantopus mexicanus, Black-necked Stilt

LATIN IN ACTION

Dohrn's Thrush-Babbler (*Horizorhinus dohrni*) is one of several birds whose ranges are so restricted and isolated that not much is known about them. Also called the Principe Flycatcher-babbler, it is restricted to the small islands of São Tomé and Principe off the west coast of Guinea. The only member of its genus *Horizorhinus* (horizontal bill), recent DNA information puts it in the genus *Sylvia* (inhabiting the woods), the Old World Warblers.

Horizorhinus dohrni,
Dohrn's Thrush-Babbler

Hoffmanni, -ii HOF-man-nye/hof-MAN-nee-eye
After Karl Hoffmann, German naturalist, as in *Pyrrhura hoffmanni*, the Sulphur-winged Parakeet

Holochlora, -us hol-o-KLOR-a/us
Greek, *holo*, whole, and *chlor*, green, as in *Psittacara holochlorus*, the Green Parakeet

Holosericeus hol-o-ser-ISS-ee-us
Greek, *holo*, whole, and *seric*, silken, as in *Eulampis holosericeus*, the Green-throated Carib, which is silky smooth and iridescent over much of the body

Homochroa, -us ho-mo-KRO-a/us
Greek, *homo*, like, same, and *chroa*, the skin, as in *Oceanodroma homochroa*, the Ashy Storm Petrel, which is ashy gray all over

Horizorhinus hor-ih-zo-RINE-us
Greek, *horiz*, horizon, and *rhinos*, bill, as in *Horizorhinus dohrni*, Dohrn's Thrush-Babbler (see box)

Hornemanni HOR-ne-man-nye
After Jens Hornemann, a Danish botanist, as in *Acanthis hornemanni*, the Arctic Redpoll

Horus HOR-us
Egyptian sun god, as in *Apus horus*, the Horus Swift, so named probably because it flies high against the bright sky.

Hottentottus hot-ten-TOT-tus
After the indigenous Khoi Khoi people of southern Africa, as in *Dicrurus hottentottus*, the Hair-crested Drongo

Hudsonia hud-SONE-ee-a
After Hudson's Bay, Canada, as in *Pica hudsonia*, the Black-billed Magpie

Hudsonicus, -a hud-SON-ih-kus/ka
After Hudson's Bay, Canada, as in *Poecile hudsonicus*, the Boreal Chickadee

Humboldti HUM-bolt-eye
After Baron Alexander von Humboldt, Prussian naturalist and explorer, as in *Spheniscus humboldti*, the Humboldt Penguin

Humei HEWM-eye
After Allan Hume, writer on Indian birds, as in *Phylloscopus humei*, Hume's Leaf Warbler

Humeralis hoo-mer-AL-is
Of the shoulder, as in *Agelaius humeralis*, the Tawny-shouldered Blackbird, referring to the colored epaulet

Humilis hoo-MIL-is
Lowly, as in *Eupodotis humilis*, the Little Brown Bustard, which flies infrequently

Hunteri HUN-ter-eye
After H. C. V. Hunter, English zoologist and big game hunter, as in *Cisticola hunteri*, Hunter's Cisticola

Huttoni HUT-ton-eye
After William Hutton, collector, as in *Vireo huttoni*, Hutton's Vireo

Hybrida hy-BRID-a
Hybrid, as in *Chlidonias hybrida*, the Whiskered Tern, hybrid probably referring to the variety of plumage and sizes in the various geographic races of the species

Hydranassa hy-dra-NASS-sa
Greek, *hydro*, water, and *anassa*, queen, as in *Hydranassa* (now *Egretta*) *tricolor*, the Tricolored Heron

Hydrobates *hy-ro-BA-teez*
Greek, *hydro*, water, and *bates*, one that walks or hunts, as in *Hydrobates pelagicus*, the European Storm Petrel

Hydrocharis *hy-dro-KAR-is*
Greek, *hydro*, water, and *charis*, favor, grace, as in *Tanysiptera hydrocharis*, the Little Paradise Kingfisher

Hydrophasianus *hy-dro-fas-ee-AN-us*
Greek, *hydro*, water, and Latin, *phasianus*, pheasant, as in *Hydrophasianus chirurgus*, the Pheasant-tailed Jacana, a water bird

Hydroprogne *hy-dro-PROG-nee*
Greek, *hydro*, water, and Latin, *progne*, swallow, as in *Hydroprogne caspia*, the Caspian Tern

Hydropsalis *hy-drop-SAL-is*
Greek, *hydro*, water, and *psalis*, scissors, as in *Hydropsalis torquata*, the Scissor-tailed Nightjar, which feeds over flooded grasslands during the tropical rainy season

Hyemalis *hy-eh-MAL-is*
Hiems, winter, and meaning wintry, as in *Junco hyemalis*, the Dark-eyed Junco, which nests in the far north of North America

Hylocharis *hy-lo-KAR-is*
Greek, *hyle*, woods, and *charis*, favor, grace, as in *Hylocharis* (now *Basilinna*) *leucotis*, the White-eared Hummingbird

Hylocichla *hy-lo-SICK-la*
Greek, *hyle*, woods, and *kichle*, thrush, as in *Hylocichla mustelina*, the Wood Thrush

Hylocryptus *hy-lo-KRIP-tus*
Greek, *hyle*, woods, and *crypt-*, hidden, as in *Hylocryptus rectirostris*, the Henna-capped Foliage-gleaner; presumably so-named because it is brownish and therefore hard to see

Hylonympha *hy-lo-NIM-fa*
Greek, *hyle*, woods, and *nympha*, nymph, as in *Hylonympha macrocerca*, the Scissor-tailed Hummingbird

Hyperborea, -us *hy-per-BOR-ee-a/us*
Greek, *hyper*, over, above, and *bore*, north, northern, as in *Larus hyperboreus*, the Glaucous Gull, referring to its home range

Hyperythra, -thrus *hy-per-IH-thra/thrus*
Greek, *hyper*, over, above, and *erythros*, red, as in *Ficedula hyperythra*, the Snowy-browed Flycatcher, referring to its very red (orange) chest

Hylonympha macrocerca,
Scissor-tailed Hummingbird

Hypocnemius *hy-pok-NEM-ee-us*
Greek, *hyper*, over, above, and *cnemi-*, lower leg, as in *Hypocnemius peruviana*, the Peruvian Warbling Antbird, whose legs seem long for its short tail

Hypocondria *hy-po-KON-dree-a*
Greek, *hyper*, over, above, and *khondros*, cartilage (of the breastbone), as in *Poospiza hypocondria*, the Rufous-sided Warbling Finch; refers to its rufous-sided chest

Hypogrammica *hy-po-GRAM-mi-ka*
Greek, *hyper*, over, above, and *grammikos*, lined, letter, as in *Pytilia hypogrammica*, the Yellow-winged Pytilia, with a lined/barred underside

Hypoleuca, -us *hy-po-LOY-ka/kus*
Greek, *hypo*, less than, and *leukos*, white, as in *Synthliboramphus hypoleucus*, the Guadalupe (once Xantus's) Murrelet, as it has less white than the Marbled Murrelet

Hypositta *hy-po-SIT-ta*
Greek, *hypo*, less than, and *sitt-a*, nuthatch, as in *Hypositta corallirostris*, the Nuthatch Vanga

Hypoxantha, -us *hy-poks-ANTH-a/us*
Greek, *hypo*, less than, and *xanth*, yellow, as in *Chelidorhynx hypoxantha*, the Yellow-bellied Fantail

I

Ianthinogaster *eye-an-thin-o-GAS-ter*
Greek, *ianthin-*, violet, and *gaster*, stomach, as in *Uraeginthus ianthinogaster*, the Purple Grenadier

Ibericus *eye-BER-ih-kus*
After Iberia (Spain and Portugal), as in *Phylloscopus ibericus*, the Iberian Chiffchaff

Ibidorhyncha *eye-bid-o-RINK-a*
Greek, *ibidos*, ibis, and *rhynch-*, bill, as in *Ibidorhyncha struthersii*, the Ibisbill

Ibis *EYE-bis*
Greek, *ibis*, stork-like bird, as in *Mycteria ibis*, the Yellow-billed Stork

Ibycter *eye-BICK-ter*
Greek, *ibu*, shout, *ibukter*, singer, as in *Ibycter americanus*, the Red-throated Caracara, with a loud and distinct call

Ichthyaetus *ik-thee-EE-tus*
Greek, *icthy*, fish, and *aetus*, eagle, as in *Ichthyaetus melanocephalus*, the Mediterranean Gull

Ichthyophaga *ik-thee-o-FAY-ga*
Greek, *icthy*, fish, and *phagein*, eat, as in *Ichthyophaga* (now *Haliaeetus*) *humilis*, the Lesser Fish Eagle

Icteria *ik-TER-ee-a*
Greek, *ikteros*, yellow, as in *Icteria virens*, the Yellow-breasted Chat

Icterina, -us *ik-ter-EE-na/nus*
Greek, *ikteros*, yellow, as in *Hippolais icterina*, the Icterine Warbler, a pale yellow bird

Icterocephala, -us *ik-ter-o-se-FAL-a/us*
Greek, *ikteros*, yellow, and Latin, *cephala*, head, as in *Tangara icterocephala*, the Silver-throated Tanager, with a yellow head

Icterophrys *ik-ter-O-friss*
Greek, *ikteros*, yellow, and *oprys*, eyebrow, as in *Satrapa icterophrys*, the Yellow-browed Tyrant

Icteropygialis *ik-ter-o-pij-ee-AL-is*
Greek, *ikteros*, yellow, and *puge*, rump, as in *Eremomela icteropygialis*, the Yellow-bellied Eremomela

Icterorhynchus *ik-ter-o-RINK-us*
Greek, *ikteros*, yellow, and Latin, *rhynchus*, bill, as in *Otus icterorhynchus*, the Sandy Scops Owl, with a yellow bill

Icterotis *ik-ter-O-tis*
Greek, *ikteros*, yellow, and *otid*, ear, as in *Ognorhynchus icterotis*, the Yellow-eared Parrot

Icterus *IK-ter-us*
Greek, *ikteros*, yellow, as in *Icterus spurius*, the Orchard Oriole; myth has it that the sight of an oriole would cure jaundice

Ictinaetus *ik-tin-EE-tus*
Greek, *iktinos*, a kite, and *aetus*, an eagle, as in *Ictinaetus malaiensis*, the Black Eagle

Icterus spurius, Orchard Oriole

LATIN IN ACTION

The Short-tailed Finch is a bird with a short tail that resembles a starling, as described by its scientific name, *Idiopsar brachyurus*. It is the only one in its genus and one of those species with a limited distribution, in this case the high Andes of Peru, Bolivia, and Argentina, at elevations of 10,800 to 15,000 feet (3,300 to 4,600 meters), in a treeless, rocky steppe habitat. Now a member of the family Emberizidae (the buntings and sparrows), it was once considered part of the blackbird family (Icteridae). This debate has been going on since 1886, when the bird was first described.

Idiopsar brachyurus, Short-tailed Finch

Ictinia *ik-TIN-ee-a*
Greek, *iktinos*, a kite, as in *Ictinia plumbea*, the Plumbeous Kite; bird kites were named after children's toy kites because of the way they fly

Idiopsar *id-ee-OP-sar*
Greek, *idio*, peculiar, and *psar*, speckled or starling, as in *Idiopsar brachyurus*, the Short-tailed Finch (see box), which vaguely resembles a starling

Ifrita *eye-FRIT-a*
After *ifrit*, supernatural creatures of fire in Arabic and Islamic culture, as in *Ifrita kowaldi*, the Blue-capped Ifrit. This genus is one of only three to contain poisonous birds

Igneus *IG-nee-us*
Fiery, as in *Pericrocotus igneus*, the Fiery Minivet

Ignicapilla *ig-ni-ka-PIL-la*
Ignis, fire, and *capilla*, hair, as in *Regulus ignicapilla*, the Common Firecrest

Ignicauda *ig-ni-KAW-da*
Ignis, fire, and *cauda*, tail, as in *Aethopyga ignicauda*, the Fire-tailed Sunbird

Ignipectus *ig-ni-PEK-tus*
Ignis, fire, and *pectus*, breast, as in *Dicaeum ignipectus*, the Fire-breasted Flowerpecker

Ignobilis *ig-NO-bil-is*
Undistinguished, of low birth, obscure, as in *Turdus ignobilis*, the Black-billed Thrush, a plain-colored bird

Iheringi *EER-ing-eye*
After Hermann von Ihering, a German ornithologist, as in *Myrmotherula iheringi*, Ihering's Antwren

Ijimae *ee-JEE-mee*
After I. Ijima, first president of the Ornithological Society of Japan, as in *Phylloscopus ijimae*, the Ijima's Leaf Warbler

Iliaca, -us *il-ee-AK-a/us*
After *ilia-*, the flank, loin, as in *Passerella iliaca*, the Fox Sparrow, so named because its most northern population is fox-colored; *iliaca* refers to the heavily streaked flanks

Ilicura *il-ih-KOO-ra*
Greek, *helix*, curl, twist, and *oura*, tail, as in *Ilicura militaris*, the Pin-tailed Manakin, named for its central pintail feathers and its plumage, which resembles a military uniform

Illadopsis *il-la-DOP-sis*
Greek, *illis*, thrush, and *opsis*, appearance, as in *Illadopsis cleaveri*, the Blackcap Illadopsis

Immaculata, -us *im-mak-oo-LAT-a/us*
Immaculate, as in *Myrmeciza immaculata*, the Blue-lored Antbird, immaculate perhaps because of its uniform plumage

Immer *IM-mer*
Immersus, dive, plunge into, as in *Gavia immer*, the Great Northern Loon or Diver

Immutabilis *im-moo-TA-bil-is*
Unchanging, as in *Phoebastria immutabilis*, the Laysan Albatross; juveniles are very similar to adults, hence the name

Impennis *im-PEN-nis*
Featherless, as in *Pinguinus impennis*, the now extinct Great Auk; although not featherless, the bird's feathers were not for flying

Imperialis *im-per-ee-AL-is*
Having a commanding presence, as in *Amazona imperialis*, the Imperial Amazon, a spectacular green and purple parrot

Implicata *im-pli-KAT-a*
Implicatus, to involve, implicate, entwine, as in *Pachycephala implicata*, the Hooded Whistler

Importunus *im-por-TOON-us*
Inconvenient, annoying, persistent, as in *Andropadus importunus*, the Sombre Greenbul; perhaps named after its monotonous whistle-like call

Inca *INK-a*
After the Inca Empire, as in *Columbina inca*, the Inca Dove; although named after the Incas, it does not occur in the Incan area of South America

Incanum, -us, -a *in-KAN-um/us/a*
Gray, as in *Tringa incana*, the Wandering Tattler, with a gray back

Incertus *in-SERT-us*
Uncertain, as in *Pseudorectes incertus*, the White-bellied Pitohui; very little is known about this bird

Incognita *in-kog-NEE-ta*
Disguised, concealed, as in *Megalaima incognita*, the Moustached Barbet

Indica *IN-di-ka*
After India, as in *Chalcophaps indica*, the Common Emerald Dove

Indicator *in-di-KA-tor*
That which points out, indicates, as in *Indicator indicator*, the Greater Honeyguide. These birds eat beeswax and lead humans to beehives to expose the hive

Indicus *IN-di-kus*
After India, as in *Caprimulgus indicus*, the Jungle Nightjar, whose common name comes from its long "night-jrrrrrrrrrrrrr" call

Indigo *IN-di-go*
Indicum, the color indigo, as in *Eumyias indigo*, the Indigo Flycatcher

Indistincta *in-dis-TINK-ta*
Indistinct, obscure, as in *Lichmera indistincta*, the Brown Honeyeater

Pinguinus impennis,
Great Auk

Tityra inquisitor,
Black-crowned Tityra

Indus *IN-dus*
India, as in *Haliastur indus*, the Brahminy Kite or Red-backed Sea Eagle

Inepta *in-EP-ta*
Ineptus, silly, foolish, as in *Megacrex inepta*, the New Guinea Flightless Rail, named for its inability to fly away from threats

Inexpectata *in-eks-pek-TA-ta*
Unexpected, as in *Pterodroma inexpectata*, the Mottled Petrel, native to New Zealand but occurs unexpectedly in other areas

Infelix *in-FEL-liks*
Unhappy, unfortunate, as in *Symposiachrus infelix*, the Manus Monarch. The type specimen upon which the species was named is said to have been in a sad condition, having been badly shot

Infuscata, -us *in-foos-KAT-a/us*
Dusky, darkened, as in *Aerodramus infuscatus*, the Halmahera Swiftlet

Ingens *IN-jenz*
Large, remarkable, as in *Megascops ingens*, the Rufescent Screech Owl, one of the larger species of screech owls

Inornatus, -a *in-or-NAT-us/a*
Without adornments, plain, as in *Baeolophus inornatus*, the Oak Titmouse, formerly the Plain Titmouse

Inquieta, -ius *in-kwee-EH-ta/ee-us*
Restless, agitated, as in *Myiagra inquieta*, the Restless Flycatcher

Inquisitor *in-KWI-zi-tor*
Inquisitor, investigator, as in *Tityra inquisitor*, the Black-crowned Tityra; probably from their head movements when foraging

Insignis *in-SIG-nis*
Conspicuous, eminent, as in *Ardea insignis*, the White-bellied Heron or Imperial Heron

Insularis *in-soo-LAR-is*
Insula, island, as in *Passer insularis*, the Socotra Sparrow of three islands in the Indian Ocean

Intermedia *in-ter-MEE-dee-a*
Intermedius, intermediate, as in *Egretta intermedia*, the Intermediate Egret, a medium-sized heron

Internigrans *in-ter-NYE-granz*
Inter, between, among, and *nig*, dark, black, as in *Perisoreus internigrans*, the Sichuan Jay, which is various shades of black and gray

Interpres *IN-ter-press*
Inter, between, and *pre-*, before, go-between, broker, as in *Arenaria interpres*, the Ruddy Turnstone, for its habit of turning over stones

Involucris *in-vo-LOO-kris*
Involucre, to wrap, as in *Ixobrychus involucris*, the Striped-backed Bittern

Iodopleura *eye-o-doe-PLUR-a*
Greek, *iodo*, violet, and *pleura*, side, as in *Iodopleura pipra*, the Buff-throated Purpletuft

Ixobrychus involucris,
Striped-backed Bittern

Iole *eye-O-lee*
Greek, mythical daughter of Eurytus, as in *Iole virescens*, the Olive Bulbul, the common name deriving from Persian for nightingale

Iphis *EYE-fiss*
Greek, mightily, strongly, as in *Pomarea iphis*, the Iphis Monarch

Irania *ee-RAHN-ee-a*
After Iran, as in *Irania gutturalis*, the White-throated Robin (actually an Old World flycatcher)

Irena *ee-REN-a*
Greek goddess of peace, as in *Irena puella*, the Asian Fairy-bluebird

Iriditorques *ih-rid-ih-TOR-kweez*
Iris, rainbow, and *torques*, collar, as in *Columba iriditorques*, the Western Bronze-naped Pigeon

Iridophanes *ih-rid-o-FAN-eez*
Iris, rainbow, and Greek, *phane*, visible, as in *Iridophanes pulcherrimus*, the Golden-collared Honeycreeper

Iridoprocne *ih-rid-o-PROK-nee*
Iris, rainbow, and Procne, who in Greek mythology was turned into a swallow by the gods, as in *Iridoprocne* (now *Tachycineta*) *bicolor*, the Tree Swallow

Iridosornis *ih-rid-o-SOR-nis*
Iris, rainbow, and *ornis*, bird, as in *Iridosornis rufivertex*, the Golden-crowned Tanager

Iris *EYE-ris*
Rainbow, as in *Pitta iris*, the Rainbow Pitta

Isabellae *ih-sa-BEL-lee*
After Queen Isabel of Spain, as in *Oriolus isabellae*, the Isabela Oriole

Isidori *iz-ih-DOR-eye*
After Isidore Geoffroy St. Hilaire, French zoologist and collector, as in *Spizaetus isidori*, Black-and-chestnut Eagle

Islandica *iss-LAN-dik-a*
After Iceland, as in *Bucephala islandica*, Barrow's Goldeneye

Ispidina *iss-pi-DEEN-a*
From *hispidus*, rough, shaggy, hairy, as in *Ispidina picta*, the African Pygmy Kingfisher

---

LATIN IN ACTION

The Sunda Bulbul (*Ixos virescens*) lives in Sumatra and Java, in Indonesia. Sunda comes from the name of the strait connecting the Java Sea to the Indian Ocean. Bulbul derives from the Persian *bolbol*, and means nightingale, although the bulbul does not belong to the nightingale family but to Pycnonotidae, the bulbuls and greenbuls. Named after mistletoe, it also eats a variety of fruits as well as insects, spiders, and other arthropods. A gregarious species, the Sunda Bulbul often forages in flocks of three to six birds of its own species as well as mixed species flocks, which it seems to prefer.

Ixos virescens, Sunda Bulbul

Ixobrychus *iks-o-BRICK-us*
Greek, *iksos*, and *brykein*, to devour, as in *Ixobrychus exilis*, the Least Bittern

Ixoreus *iks-OR-ee-us*
Greek, *iksos*, mistletoe, and *oro*, a mountain, as in *Ixoreus naevius*, the Varied Thrush, referring to its preferred mountain habitat and diet

Ixos *IKS-os*
Greek, *iksos*, mistletoe, as in *Ixos virescens*, the Sunda Bulbul (see box)

James Bond
(1900–1989)

Bird watchers may be surprised to discover that the original, real, James Bond, after whom Ian Fleming's fictional character was named, was one of their own. Bond was born January 4, 1900, in Philadelphia, but when his mother died in 1914, he moved to England with his British-born father. There he went to private school and then to Cambridge University, completing his degree in 1922.

His interest in ornithology was sparked by his ornithologist father's expedition to the Orinoco Delta right after graduation. He returned to the US, and spent three years working as a banker, but his interest in natural history led him to take on a role in an expedition sponsored by the Academy of Natural Sciences, which involved surveying the birds of the West Indies. He travelled extensively through the islands for many decades, spending long periods in Cuba and Hispaniola. "Virtually the entire area was explored fairly thoroughly with the exception of some of the more southern Bahamas," he wrote in 1960. "Of the native West Indian species of birds and those known to have been successfully introduced I encountered approximately 98 percent in life."

He led a series of trips to document avian species throughout the Caribbean. One island that fascinated him was Jamaica, where he noticed that many of the bird species native to that island originated from North America, and not South America as had been originally assumed. Later trips to Jamaica and other Caribbean islands led him to the theory that the boundary between North and South American species lay off the northeast coast of Venezuela and Columbia, now called the Bond Line. Bond wrote up the seminal book of Caribbean bird-watching, *Birds of the West Indies*, originally published in 1936 and for many years the only definitive bird identification book of the area. He visited more than 100 islands and collected 294 of the 300 bird species there, often while plying the water around the islands in makeshift canoes. He ultimately wrote more than 100 scientific papers on Caribbean birds.

His *Birds of the West Indies* was widely read by bird-watchers in the Caribbean area. One such bird-watcher, Ian Fleming, had an estate on the north coast of Jamaica and used Bond's book as a guide for his birding forays. His selection of Bond's name for the hero of his spy novels made the name, if not

Todus todus,
Jamaican Tody

The Jamaican Tody is endemic to Jamaica; *todillus* is Latin for small bird.

"The country seems very large!"

James Bond, aged 53. He made this remark en route to Michigan, travelling west of Philadelphia for the first time.

the man, famous. Fleming chose the name because he liked its strength and simplicity, and figured that the real Bond had no objections, although he was not asked. Bond did not even notice for several years.

The popularity of Fleming's books eventually caused some consternation to the ornithologist. Bond's wife Mary wrote jokingly to Fleming that she was appalled that in the novel *Dr. No*, the wily rascal was named James Bond. In response, Fleming said that James could sue if he wished or "Perhaps one day he will discover some particularly horrible species of bird which he would like to christen in an insulting fashion." Interestingly, Fleming used a bird sanctuary on Crab Key on Inagua Island in the Bahamas as the setting for *Dr. No*.

In 1964 James and his wife Mary were in the Caribbean to continue research on bird species and decided to pay a surprise visit to Ian Fleming, who had on the first exchange of letters invited them to his estate in Jamaica. Fleming was very ill, with about six months to live. By chance the BBC was doing an interview with Fleming, who had become almost as famous as the fictional James Bond, so they were able to film the one and only meeting between these two authors. At first Fleming was somewhat suspicious, asking Bond to identify some of the birds they saw on the premises. But once Bond passed the test, this was probably the best day Fleming would have for the rest of his life.

During a full professional life, Bond was a curator at the Academy of Natural Sciences of Philadelphia, a fellow of the American Ornithologists' Union, and a member of the British Ornithologists' Union. In 1952 he received the Musgrave Medal from the Institute of Jamaica, and in 1954 he was awarded the William Brewster Memorial Award, the most prestigious accolade in American ornithology, by the American Ornithologists' Union for his work on West Indian birds, and the Leidy Medal of the Academy of Natural Sciences in 1975.

He died in Philadelphia at age 89.

Ardea herodias,
Great Blue Heron

A Great Blue Heron in Galapagos National Park has been nicknamed James Bond because its band/ring number is 007.

J

Jabiru *ja-BEER-oo*
From Tupi (indigenous to Brazil), swollen neck, as in *Jabiru mycteria*, the Jabiru; the head and upper neck are naked and black, with a naked leather-like red expandable pouch at the base

Jacamaralcyon *jak-a-mar-AL-see-on*
Jacamar, from Tupi (indigenous to Brazil), and Greek, *alkuon*, kingfisher, as in *Jacamaralcyon tridactyla*, the Three-toed Jacamar

Jacamerops *ja-ka-MER-ops*
Jacamar, from Tupi (indigenous to Brazil), and *merops*, bee, as in *Jacamerops aureus*, the Great Jacamar

Jacana *ja-KA-na*
Tupi-Guanari language, as in *Jacana spinosa*, the Northern Jacana

Jacarina *ja-ka-REEN-a*
Tupi name for one who jumps up and down, as in *Volatinia jacarina*, the Blue-black Grassquit, the male of which jumps into the air while singing

Buteo jamaicensis, Red-tailed Hawk

Jacksoni *JAK-son-eye*
After Frederick Jackson, English administrator, naturalist, and ornithologist, as in *Tockus jacksoni*, Jackson's Hornbill

Jacobinus *ja-ko-BINE-us*
After Dominican friars or Jacobins, as in *Clamator jacobinus*, the Jacobin or Pied Cuckoo; both the friars and the birds are white with a black "cloak"

Jacquinoti *jak-kwee-NOTE-eye*
After Charles Jacquinot, French explorer, as in *Ninox jacquinoti*, the Solomons Boobook

Jacucaca *ja-koo-KA-ka*
Tupi name, as in *Penelope jacucaca*, the White-browed Guan

Jacula *ja-KOO-la*
Jacul-, throw, as in *Heliodoxa jacula*, the Green-crowned Brilliant; large for a hummingbird, it feeds while perched but darts to other perches

Jamaicensis *ja-may-SEN-sis*
After Jamaica, as in *Buteo jamaicensis*, the Red tailed Hawk

Jambu *JAM-boo*
Sanskrit, rose-apple tree, as in *Ptilinopus jambu*, the Jambu Fruit Dove

Jamesi *JAMEZ-eye*
After Henry James, a British businessman, as in *Phoenicoparrus jamesi*, James's Flamingo

Jamesoni *JAY-meh-son-eye*
After James Jameson, Irish hunter and naturalist, as in *Platysteira jamesoni*, Jameson's Wattle-eye

Jankowskii *jan-KOW-skee-eye*
After Michael Jankowski, Polish zoologist, as in *Emberiza jankowskii*, Jankowski's Bunting

Janthina *jan-THEEN-a*
Greek, *ianthinos*, violet-colored, as in *Columba janthina*, the Japanese Wood Pigeon

Japonica, -us, *ja-PON-ik-a/us*
Of Japan, as in *Zosterops japonicus*, the Japanese White-eye

Jardineii, -i *jar-DINE-ee-eye/jar-DINE-ee*
After William Jardine, Scottish ornithologist, as in *Turdoides jardineii*, the Arrow-marked Babbler

Javanica, -us *ja-VAN-ih-ka/kus*
Of Java, as in *Rhipidura javanica*, the Malaysian Pied Fantail

Jelskii *JEL-skee-eye*
After Konstanty Jelski, Polish ornithologist, as in *Silvicultrix jelskii*, Jelski's Chat-Tyrant

Jerdoni *JER-don-eye*
After Thomas Jerdon, British physician and naturalist, as in *Aviceda jerdoni*, Jerdon's Baza

Jocosus *jo-KO-sus*
Full of fun, as in *Pycnonotus jocosus*, the Red-whiskered Bulbul

Johannae *jo-HAN-nee*
After Johanna Verreaux, wife of Jules Verreaux, as in *Cinnyris johannae*, Johanna's Sunbird

Jefferyi *JEF-free-eye*
After the father of John Whitehead, English explorer and naturalist, and professional collector, Jeffery Whitehead, as in *Pithecophaga jefferyi*, the Phillipine Eagle

Johnstoni *JON-stun-eye*
After Harry Johnston, English explorer and administrator, as in *Ruwenzorornis johnstoni*, Ruwenzori Turaco

Johnstoniae *jon-STONE-ee-eye*
After Marion Johnstone, famous aviculturist, as in *Tarsiger johnstoniae*, the Collared Bush Robin

Jonquillaceus *jon-kwil-LACE-ee-us*
French, narcissus, as in *Aprosmictus jonquillaceus*, the Jonquil Parrot; perhaps because of the yellowish-olive wing coverts that resemble the yellow of narcissus

Josefinae/Josephinae *jo-seh-FIN-ee*
After the wife of German ornithologist Friedrich Finsch, as in *Charmosyna josefinae*, Josephine's Lorikeet; and as in *Hemitriccus josephinae*, the Boat-billed Tody-Tyrant

Jourdanii *joor-DAN-ee-eye*
After a collector in Trinidad, as in *Chaetocercus jourdanii*, the Rufous-shafted Woodstar

Jouyi *JOO-ee-eye*
After Pierre Jouy, American diplomat and naturalist, as in *Columba jouyi*, the extinct Ryukyu Wood Pigeon

Jynx ruficollis, Red-throated Wryneck

Jubata, -us, -ula *joo-BAT-a/us/joo-ba-TOO-la*
Jubatus, a crest or mane, as in *Chenonetta jubata*, the Australian Wood Duck/Maned Duck

Jugularis *jug-oo-LAR-is*
Jugularis, of the collarbone, throat, neck, as in *Brotogeris jugularis*, the Orange-chinned Parakeet

Julie *JOO-lee*
After Julie Mulsant, wife of the French naturalist, Martial Mulsant, as in *Damophila julie*, the Violet-bellied Hummingbird

Juncidis *jun-SID-is*
Juncus, rush, as in *Cisticola juncidis*, the Zitting Cisticola or Fan-tailed Warbler, found in grasslands, often near water

Junco *JUNK-o*
Juncus, rush, as in *Junco hyemalis*, the Dark-eyed Junco. An odd generic name as they are not wetland birds

Jynx *JINKS*
Wryneck, as in *Jynx ruficollis*, the Red-throated Wryneck, with a very flexible neck

K

Kaempferi *KEMP-fer-eye*
After Emil Kaempfer, German collector, as in *Hemitriccus kaempferi*, Kaempfer's Tody-Tyrant

Kaestneri *KEST-ner-eye*
After Peter Kaestner, American diplomat, as in *Grallaria kaestneri*, the Cundinamarca Antpitta, after Cundinamarca, Colombia

Kakamega *ka-ka-MAY-ga*
After the Kakamega Rainforest in Kenya, as in *Kakamega poliothorax*, the Gray-chested Babbler

Kansuensis *kan-su-EN-sis*
After Kansu/Gansu Province, China, as in *Phylloscopus kansuensis*, the Gansu Leaf Warbler

Kandti *KANT-eye*
After Richard Kandt, German physician and explorer, as in *Estrilda kandti*, Kandt's Waxbill

Kaupifalco *kaw-pi-FAL-ko*
After Johann Kaup and *falco*, falcon, as in *Kaupifalco monogrammicus*, the Lizard Buzzard

Kawalli *KA-wal-lye*
After Nelson Kawall, Brazilian aviculturist, as in *Amazona kawalli*, Kawall's, or White-faced, Amazon

Kelleyi *KEL-lee-eye*
After W. V. Kelley, American philanthropist, as in *Macronus kelleyi*, the Gray-faced Tit-Babbler

Kempi *KEMP-eye*
After Robert Kemp, American naturalist and collector, as in *Macrosphenus kempi*, Kemp's Longbill

Kennicotti *KEN-ih-kot-tye*
After Robert Kennicott, American Naturalist, as in *Megascops kennicotti*, the Western Screech Owl

Kenricki *KEN-rik-eye*
After R. W. E. Kenrick, British Army officer, as in *Poeoptera kenricki*, Kenrick's Starling

Macrosphenus kempi, Kemp's Longbill

Keraudrenii *ke-raw-DREN-ee-eye*
After Pierre Keraudren, French physician, as in *Phonygammus keraudrenii*, the Trumpet Manucode

Ketupu *ke-TOO-poo*
Malay name for bird, as in *Ketupa ketupu*, the Buffy Fish Owl

Kienerii, -i *kee-NAIR-ee-eye/KEEN-er-eye*
After Louis-Charles Kiener, French malacologist (study of molluscs), as in *Lophotriorchis kienerii*, the Rufous-bellied Hawk-Eagle

Kilimensis *ki-li-MEN-sis*
After Mt. Kilimanjaro, Tanzania, as in *Nectarinia kilimensis*, the Bronzy Sunbird

Kirhocephalus *keer-ho-se-FAL-us*
Greek, *kirrhos*, tawny, orange-colored, and Latin, *cephala*, head, as in *Pitohui kirhocephalus*, the Northern Variable Pitohui, with a mostly orange body and black head

Kirki *KIRK-eye*
After John Kirk, Scottish physician and administrator, as in *Zosterops kirki*, Kirk's White-eye

Kirtlandii *kirt-LAN-dee-eye*
After Jared Kirtland, American doctor, naturalist, and botanist, as in *Setophaga kirtlandii*, Kirtland's Warbler

Klaas *KLAAS*
After a famous servant who apparently discovered the bird, as in *Chrysococcyx klaas*, Klaas's Cuckoo

Klagesi KLAIGS-eye
After Samuel Klages, American collector, as in *Myrmotherula klagesi*, Klages's Antwren

Knipolegus ni-po-LAY-gus
Greek, *knipos*, insect, and *legus*, choose, as in *Knipolegus signatus*, the Andean Tyrant

Kochi KOCK-eye
After Gottleib von Koch, German collector and taxidermist, as in *Erythropitta kochi*, the Whiskered Pitta

Koepckeae KEP-kee-ee
After Maria Koepcke, Mother of Peruvian ornithology, as in *Cacicus koepckeae*, the Selva Cacique

Komadori kom-a-DOR-eye
Japanese for Red Robin, as in *Erithacus komadori*, the Ryukyu Robin

---

LATIN IN ACTION

The Whiskered Pitta, *Erythropitta kochi*, is an unusually beautiful bird, with a bright-red lower chest and abdomen, and topped by an iridescent upper chest and throat, a brown head, and greenish back. Pitta is from a language of southern India and parts of Sri Lanka and means pretty bauble; certainly very descriptive of this bird.

Erythropitta kochi, Whiskered Pitta

Sitta krueperi, Krüper's Nuthatch

Kona KO-na
From the Hawaiian Islands, as in *Chloridops kona*, the Kona Grosbeak

Kori KOR-eye
From Setswana (South African language) *kgori*, as in *Ardeotis kori*, the Kori Bustard; Bustard may have come from the Latin *aves tarda*, slow bird

Kozlowi KOZ-low-eye
After Pyotr Kozlov, Russian explorer, as in *Prunella koslowi*, the Mongolian Accentor

Kretschmeri KRETCH-mer-eye
After Eugen Kretschmer, a German collector, as in *Macrosphenus kretschmeri*, Kretschmer's Longbill

Krueperi KRUE-per-eye
After Theobald Krüper, German ornithologist, as in *Sitta krueperi*, Krüper's Nuthatch

Kubaryi koo-BARY-eye
After Jan Kubary, Polish explorer, as in *Corvus kubaryi*, the Mariana Crow

Kuehni KOON-eye
After Heinrich Kühn, German naturalist, as in *Myzomela kuehni*, the Crimson-hooded Myzomela

Kupeornis koo-pee-OR-nis
From Mt. Kupe in the Cameroon, and Greek, *ornis*, bird, as in *Kupeornis gilberti*, the White-throated Mountain Babbler

Feathers

Like mammals, birds are homeothermic (warm-blooded); like many reptiles, amphibians, fish, and a couple of mammals, they lay eggs. They show parental care, as do mammals and some reptiles and fish; they migrate, as do some mammals and fish. But unlike other animal groups, birds are very recognizable because their characteristics are fairly homogeneous, and uniquely they have feathers. If an animal has feathers, it is a bird.

Often called the first bird, *Archaeopteryx* (ancient wing) *lithographica* is a creature that lived about 150 million years ago. Eleven fossils have been recovered from a limestone quarry in Germany, hence the specific epithet *lithographica*, limestone being used in making lithographs. *Archaeopteryx lithographica* was clearly an intermediate form between dinosaurs and birds as it had teeth, a long bony tail, claws on its hands, and other reptilian characteristics, but it also had well-developed feathers. Whether it could fly or just glide is the object of speculation, but flight-like feathers are present.

Primary feathers are attached to the hand primarily for propulsion; secondary feathers are attached to the forelimb primarily for lift.

Feathers first evolved not for flight but for insulation. In the changes dinosaurs experienced over millions of years, there is evidence that they were developing homeothermic capabilities—becoming warm-blooded. To do that, bodies needed something to prevent rapid heat loss. Scales and feathers are made of keratin so it is likely that scales elongated, split, and became thinner, evolving into the prototype of feathers. Only many years later did feathers elongate enough for gliding and then powered flight.

As feathers evolved, they differentiated into various forms for diverse purposes. The down feathers, as we know, perform the original feather function of insulation. Flight feathers, those of the hand, serve to propel the bird through the air (or water in the case of swimmers) in a figure eight fashion, as seen from the side. Other feathers attached to the arm provide lift like an airplane wing. The tail feathers serve both as rudder and brake. Feathers called contour feathers cover the body to make it smooth and aerodynamic. Semiplumes—feathers structurally intermediate between down and contour feathers—help both in waterproofing and giving the bird a sleek profile.

Archaeopteryx lithographica

Archaeopteryx lithographica was about the size of a raven and recent evidence indicates that the feathers were black.

All of these feathers need to be preened and oiled to avoid becoming matted or waterlogged. A "preen" gland on the top of the tail base produces oil that the bird squeezes out and runs over its feathers. Powder down, found in some birds like herons and egrets, are feathers whose ends break off into talc-like particles that along with preen gland oil help waterproof feathers.

Specialized feathers like filoplumes (those "hairs" you see on a plucked chicken) provide the bird with information about the position of its body feathers. Rictal bristles at the sides of the jaw apparently serve to tell a flying bird about its position in the air as well as its speed.

Insulation first, flight second, and then camouflage or courtship are the evolving uses of feathers. Birds have developed extremely clever ways of hiding themselves from predators by being cryptically colored. For example, many plovers disrupt their outlines with breastbands; the females of many species just use dull plumage. On the other extreme, many male birds use elaborate plumages with bright, even iridescent colors to establish a territory, attract females, and defend their nest site. Plumes, fans, bristles, crests, elongated tails, and an endless variety of patterns and colors decorate birds. The lyrebird, turkey, and peacock all display large, fancy tails. The Amazonian Royal Flycatcher, *Onychorhynchus coronatus*, can display a large, bright, fan-shaped crest when in the mood, and the Kagu, *Rhynochetos jubatus*, can raise its long head feathers, usually draped down the back of its neck.

Since feathers can comprise 20 percent of the weight of a bird, they are clearly important. Why, otherwise, would a hummingbird sprout 1,000 of them and a swan 25,000?

There are a number of Latin and Greek suffixes that refer to feathers in some way: *petryl*, *ptero*, *ptilo*, *ptin*, *pinna*, and *penna*, a feather or wing; *pinnat-*, *ptin*, feathered; *ala*, *ali-*, wing, *alat-*, *pten*, winged.

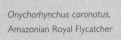

Onychorhynchus coronatus,
Amazonian Royal Flycatcher

The Royal Flycatcher male has a showy crest that it only raises during mating and preening.

Pavo cristatus,
Indian Peafowl

The "eyed" tail feathers of the peacock are shed yearly and increase in length and number with maturity.

L

Labradorius *la-bra-DOR-ee-us*
After Labrador, Canada, as in the now extinct *Camptorhynchus labradorius*, the Labrador Duck

Lactea *LAK-tee-a*
Lacte, milk, as in *Polioptila lactea*, the Creamy-bellied Gnatcatcher

Laeta *LEE-ta*
Gay, pleasing, as in *Cercomacra laeta*, Willis's Antbird

Lafayetii *la-fye-ET-eye*
After Marie du Mothier, Marquis de Lafayette, as in *Gallus lafayetii*, the Sri Lanka Junglefowl

Lafresnayi *la-FREZ-nay-eye*
After Noel Andre de La Fresne, French ornithologist and collector, as in *Picumnus lafresnayi*, Lafresnaye's Piculet

Lagdeni *LAG-den-eye*
After Godfrey Lagden, English diplomat, as in *Malaconotus lagdeni*, Lagden's Bushshrike

Lagonosticta *la-go-no-STICK-ta*
Greek, *lagonos*, flank, and *stiktos*, spotted, dotted, as in *Lagonosticta rufopicta*, the Bar-breasted Firefinch

Lagopus *la-GO-pus*
Greek, *lagos*, hare, and *pous*, foot, as in *Lagopus lagopus*, the Willow Ptarmigan, with feathery feet that help it to walk on soft snow

Lalage *la-LA-jee*
Lallo, perhaps a girl's name, as in *Lalage nigra*, the Pied Triller

Lampornis *lam-POR-nis*
Greek, *lampro*, a torch, light, and *ornis*, bird, as in *Lampornis viridipallens*, the Green-throated Mountaingem; the name probably refers to its attractive plumage

Lamprolaima *lam-pro-LAY-ma*
Greek, *lampro*, shining, and *laima*, throat, as in *Lamprolaima rhami*, the Garnet-throated Hummingbird

Lamprolia *lam-PROL-ee-a*
Greek, *lampro*, shining, as in *Lamprolia victoriae*, the Silktail, with a bright white rump that stands out like a light

Lampropsar *lam-PROP-sar*
Greek, *lampro*, shining, and *psar*, starling, as in *Lampropsar tanagrinus*, the Velvet-fronted Grackle, an iridescent bird that resembles a starling

Lamprospiza *lam-pro-SPY-za*
Greek, *lampro*, shining, and *spiza*, finch, as in *Lamprospiza melanoleuca*, the Red-billed Pied Tanager, a brightly colored tanager that somewhat resembles a finch

Lamprotornis *lam-pro-TOR-nis*
Greek, *lampro*, shining, and *ornis*, bird, as in *Lamprotornis ornatus*, the Principe Starling, a metallic-plumaged bird

Lanaiensis *lan-eye-EN-sis*
After Lanai, Hawaii, as in *Myadestes lanaiensis*, the Olomao

Lanceolata, -us *lan-see-o-LAT-a/us*
Shaped like a spear, as in *Chiroxiphia lanceolata*, the Lance-tailed Manakin, alluding to its central tail feathers

Langsdorffi *LANGZ-dorf-fye*
After Georg von Langsdorff, German physician and naturalist, as in *Discosura langsdorffi*, the Black-bellied Thorntail

Languida *lan-GWEE-da*
Weak, faint, as in *Hippolais languida*, Upcher's Warbler, perhaps because of its slow, deliberate movements

Laniarius *lan-ee-AR-ee-us*
Lanius, butcher, and *arius*, pertaining to, as in *Laniarius ruficeps*, the Red-naped Bushshrike

Gallus lafayetii, Sri Lanka Junglefowl

Lanius

The Latin word for butcher, *Lanius* (*LAN-ee-us*), is an oft-applied name for birds in the family Laniidae. The genus contains 27 species, mostly called shrikes, a name that may come from the Old English *scric*, referring to a bird with a shrill call. Some in the genus are called fiscals, after an Afrikaans word *fiskaal*, a public official, particularly a hangman. Shrikes are carnivorous birds with a hooked upper bill they use to prey on large insects and small vertebrates. They impale their prey on thorns, spines, or barbed wire, for eating later, hence the comparisons with hangmen and butchers.

Shrikes require two types of perches, one for hunting and one for evening roosting. The birds sit upright on their diurnal perch and swoop down on prey with rapid wingbeats. They are very territorial and require a variety of perch heights. In agricultural areas their territories are larger due to a restricted choice of perches and a lower density of potential prey. During the breeding season, male shrikes will store food items in a cache. One study of Northern or Great Gray Shrikes (*L. excubitor*, Latin for sentinel) discovered that the cache of impaled prey increased as the breeding season progressed, peaked when nests were built and eggs laid, and decreased when young and mates were fed. The conclusion was that males with bigger prey caches were more successful in attracting females and raising young.

Most songbirds sing only during the breeding season, but both sexes of the Northern Shrike sing for much of the year, including winter. It turns out that the Northern Shrike mimics the calls of songbirds, one of its major prey items, in order to attract them. As carnivores, shrikes eat not only birds but also a variety of invertebrates, mammals, reptiles, and amphibians, mostly smaller than the shrike but occasionally larger. Like many raptors, shrikes regurgitate pellets of indigestible material. The Red-backed Shrike, *L. collurio*, is sometimes called Nine-killer because it was once thought they killed nine animals before eating them. Its German name is *Neuntoeter* (nine killer).

Lanius mackinnoni
Mackinnon's Shrike

Lanius isabellinus,
Isabelline Shrike

Laniisoma *lan-ee-eye-SO-ma*
Lanius, butcher, and Greek, *soma*, body, as in *Laniisoma elegans*, the shrike-like Brazilian Laniisoma

Lanio *LAN-ee-o*
Lanius, butcher, as in *Lanio fulvus*, the Fulvous Shrike-Tanager

Laniocera *lan-ee-o-SER-a*
Lanius, butcher, and *cera*, wax, as in *Laniocera rufescens*, the Speckled Mourner; *cera* refers to the bill and derives from the Greek *keras*, horn, or bill, as the bill appears waxy

Lanioturdus *lan-ee-o-TUR-dus*
Lanius, butcher, and *turdus*, thrush, as in *Lanioturdus torquatus*, the White-tailed Shrike

Lanius *LAN-ee-us*
Butcher, as in *Lanius cristatus*, the Brown Shrike

Lapponica, -us *lap-PON-i-ka/kus*
Lapland, as in *Limosa lapponica*, the Bar-tailed Godwit

Larosterna *lar-o-STIR-na*
Larus, gull, and Dutch, *sterna*, tern, as in *Larosterna inca*, the Inca Tern

Larus *LA-rus*
Gull, as in *Larus pacificus*, the Pacific Gull

Larvatus, -a *lar-VA-tus/ta*
Lavare, to bewitch, enchant, also masked, as in *Coracina larvata*, the Sunda Cuckooshrike, with a gray head/hood

Lateralis *lat-er-AL-is*
Latus, side, flank, as in *Cisticola lateralis*, the Whistling Cisticola, which has rufous margins on its wing feathers that when folded show a rufous patch

Laterallus *lat-er-AL-lus*
Latus, side, flank, and *rallus*, rail, as in *Laterallus xenopterus*, the Rufous-faced Crake, with white bars on its sides

Lathami *LAY-them-eye*
After John Latham, British physician and naturalist, as in *Peliperdix lathami*, Latham's, or Forest, Francolin

Lathamus *LAY-them-us*
After John Latham, British physician and naturalist, as in *Lathamus discolor*, the Swift Parrot

Latirostris *lat-ih-ROSS-tris*
Latus, broad, and *rostris*, beak, as in *Contopus latirostris*, the Lesser Antillean Pewee

Latistriata, -us *lat-ih-stree-AT-a/us*
Latus, broad, and *striatus*, furrow, streak, as in *Zosterornis latistriatus*, the Panay Striped Babbler, after Panay, Phillipines

Latrans *LAY-tranz*
Latrare, to bark, as in *Ducula latrans*, the Barking Imperial Pigeon

Laudabilis *law-DA-bi-lis*
Praiseworthy, as in *Icterus laudabilis*, the Saint Lucia Oriole

Lawesii *lawz-ee-eye*
After William Lawes, British New Guinea missionary, as in *Parotia lawesii*, Lawes's Parotia

Lawrencei, -ii *LAW-ren-sye/law-RENS-ee-eye*
After George Lawrence, American businessman and amateur ornithologist, as in *Spinus lawrencei*, Lawrence's Goldfinch

Layardi *lay AR dye*
After Edgar Layard, Italian collector and later curator, as in *Sylvia layardi*, Layard's Warbler

Laysanensis *lay-sa-NEN-sis*
After the Laysan Islands, as in *Anas laysanensis*, the Laysan Duck or Laysan Teal

Lazuli *la-ZOO-lye*
Lazul, azure, blue, as in *Todiramphus lazuli*, the Lazuli Kingfisher

Icterus laudabilis, Saint Lucia Oriole

Leachii LEACH-ee-eye
After William Leach, a British zoologist, as in *Dacelo leachii*, the Blue-winged Kookaburra

Leadbeateri led-BEET-ter-eye
After Benjamin Leadbeater, a British taxidermist and ornithologist, as in *Lophochroa leadbeateri*, Major Mitchell's, or Leadbeater's, Cockatoo

Lecontei, -ii le-CONT-eye/ee-eye
After John LeConte, American entomologist, as in *Toxostoma lecontei*, Le Conte's Thrasher

Legatus le-GAT-us
Ambassador, envoy, as in *Legatus leucophaius*, the Piratic Flycatcher

Leiothrix lay-EYE-o-thriks
Greek, *leios*, smooth, and *thrix*, hair, as in *Leiothrix lutea*, the Red-billed Leiothrix or Pekin Nightingale; refers to the bird's smooth feathers

Leipoa lay-eye-PO-a
Greek, *leipo*, leave, and *oon*, egg, as in *Leipoa ocellata*, the Malleefowl, which constructs a compost pile to incubate its eggs

Lentiginosus len-ti-ji-NO-sus
Lentigo, freckled, as in *Botaurus lentiginosus*, the American Bittern; a reference to the bird's patterened plumage

Lepida le-PEE-da
Lepidus, neat, elegant, as in *Rhipidura lepida*, the Palau Fantail, an elegant bird

Toxostoma lecontei,
Le Conte's Thrasher

Lepidocolaptes le-pi-doe-ko-LAP-teez
Lepidus, neat, elegant, and *colaptes*, a chisel or chiseler, as in *Lepidocolaptes affinis*, the Spot-crowned Woodcreeper, an elegantly-plumaged woodcreeper

Lepidopyga le-pi-doe-PI-ga
Lepidus, neat, and *pyga*, rump, as in *Lepidopyga lilliae*, the Sapphire-bellied Hummingbird

Lepidothrix le-pih-DOE-thrix
Lepidus, scaly, and *thrix*, hair, as in *Lepidothrix coronata*, the Blue-crowned Manakin, an elegantly plumaged Mannikin

Leptasthenura lep-tas-then-OO-ra
Greek, *leptos*, slender, fine, *asthenia*, weak, and *oura*, tail, as in *Leptasthenura andicola*, the Andean Tit Spinetail

Leptocoma lep-toe-KO-ma
Greek, *leptos*, slender, fine, and *kome*, hair, as in *Leptocoma minima*, the Crimson-backed Sunbird; the back, shoulders, and chest feathers appear as fine hairs

---

LATIN IN ACTION

Lepidothrix coronata, the Blue-crowned Manakin, describes a bird with a crown of scaly hair. The male's crown is composed of iridescent blue feathers that resemble scales and a dark blue body. The much drabber female is a blend-into-the-environment dull green. Differences in male and female coloration is called "sexual dichromatism" and occurs primarily because the male wants to attract females and the female wants to be inconspicuous while attending to the nest. The name Manakin comes from the Dutch *mannekjin*, meaning little person, although it is unclear whether it is the size of the bird or its behavior that remind one of a little man. Manakins belong to the family Pipridae, consisting of 60 species, and are distinguished from similar family groups by the shape of their syrinx (voicebox).

Leptodon lep-TOE-don
Greek, *leptos*, slender, fine, and *odon*, tooth, as in *Leptodon forbesi*, the White-collared Kite, with a sharply down-curved bill tip

Leptopoecile lep-toe-poy-SIL-ee
Greek, *leptos*, slender, fine, and *poecil-*, variegated, many-colored, as in *Leptopoecile elegans*, the Crested Tit-warbler

Leptopogon lep-toe-PO-gon
Greek, *leptos*, slender, fine, and *pogon*, beard, as in *Leptopogon rufipectus*, the Rufous-breasted Flycatcher; slender beard probably refers to the rictal bristles

Leptosittaca lep-to-SIT-a-ka
Greek, *leptos*, slender, fine, and *psittaca*, parrot, as in *Leptosittaca branickii*, the Golden-plumed Parakeet

Leptopterus lep-TOP-ter-us
Greek, *leptos*, slender, fine, and *pteron*, feathers or wing, as in *Leptopterus chabert*, the Chabert Vanga; its narrow wings are almost swallow-like

Leptoptilos, -a lep-top-TIL-os/a
Greek, *leptos*, slender, fine, and *ptilon*, wing, as in *Leptoptilos crumenifer*, the Marabou Stork

Enicognathus leptorhynchus,
Slender-billed Parakeet

Leptorhynchus lep-toe-RINK-us
Greek, *leptos*, slender, fine, and *rhynchos*, bill, as in *Enicognathus leptorhynchus*, the Slender-billed Parakeet

Leptosomus lep-tow-SO-mus
Greek, *leptos*, slender, fine, and *soma*, body, as in *Leptosomus discolor*, the Cuckoo Roller; its large head may make the body appear slender

Lepturus lep-TOOR-us
Greek, *leptos*, slender, fine, and *oura*, tail, as in *Phaethon lepturus*, the White-tailed Tropicbird

Lesbia LEZ-bee-a
Lesbia was the literary pseudonym of the great love of Roman poet Gaius Valerius Catullus, as in *Lesbia victoriae*, the Black-tailed Trainbearer

Lessonia, -i, -ii les-SON-ee-a/eye/ee-eye
After Rene Lesson, French ornithologist, as in *Lessonia oreas*, the Andean Negrito

Leucocephala, -o, -us loy-ko-se-FAL-a/o/us
Greek, *leuko*, white, and Latin, *cephala*, the head, as in *Columba* (now *Patagioenas*) *leucocephala*, the White-crowned Pigeon

Leucochloris loy-ko-KLOR-is
Greek, *leuko*, white, and *chloris*, greenness, freshness, as in *Leucochloris albicollis*, the White-throated Hummingbird

Leucogaster, -ra loy-ko-GAS-ter/ra
Greek, *leuko*, white, and *gaster*, abdomen, as in *Sula leucogaster*, the Brown Booby

Leucogenys loy-ko-JEN-is
Greek, *leuko*, white, and Latin, *gena*, cheek, as in *Conirostrum leucogenys*, the White-eared Conebill

Leucolaema, -us loy-ko-LEE-ma/mus
Greek, *leuko*, white, and *laemus*, throat, as in *Geokichla leucolaema*, the Enganno Thrush, after the island of Enganno, Indonesia

Leucolophus loy-ko-LO-fus
Greek, *leuko*, white, and *lophus*, crest, tuft, as in *Tauraco leucolophus*, the White-crested Turaco

Leucomelas, -a loy-ko-MEL-as/a
Greek, *leuko*, white, and *melas*, black, dark, as in *Tricholaema leucomelas*, the Acacia Pied Barbet

Columba leuconota,
Snow Pigeon

Leuconota, -us *loy-ko-NO-ta/tus*
Greek, *leuko*, white, and *notos*, back, as in *Columba leuconota*, the Snow Pigeon

Leucopeza *loy-ko-PEH-za*
Greek, *leuko*, white, and *peza*, foot, edge, as in *Leucopeza semperi*, the possibly extinct Semper's Warbler

Leucophrys *loy-KO-fris*
Greek, *leuko*, white, and *ophyrs*, brow, eyebrow, as in *Zonotrichia leucophrys*, the White-crowned Sparrow

Leucophthalma, -us *loy-kof-THAL-ma/mus*
Greek, *leuko*, white, and *ophthalma*, eye, as in *Psittacara leucophthalmus*, the White-eyed Parakeet

Leucopleura, -us *loy-ko-PLUR-a/us*
Greek, *leuko*, white, and *pleura*, side, as in *Thescelocichla leucopleura*, the Swamp Palm Bulbul

Leucopogon *loy-ko-PO-gon*
Greek, *leuko*, white, and *pogon*, beard, as in *Cantorchilus leucopogon*, the Stripe-throated Wren

Leucopsar *loy-KOP-sar*
Greek, *leuko*, white, and *psar*, a starling, as in *Leucopsar rothschild*, the Bali Myna

Leucopsis *loy-KOP-sis*
Greek, *leuko*, white, and *opsis*, appearance, as in *Branta leucopsis*, the Barnacle Goose, so called because they were once thought to hatch from barnacles

Leucoptera, -us *loy-KOP-ter-a/us*
Greek, *leuko*, white, and *ptera*, wing, as in *Loxia leucoptera*, the Two-barred Crossbill

Leucopternis *loy-kop-TER-nis*
Greek, *leuko*, white, and *pternis*, hawk, as in *Leucopternis melanops*, the Black-faced Hawk

Leucopyga, -alis *loy-ko-PIJ-a/loy-ko-pij-AL-is*
Greek, *leuko*, white, and *puge*, rump, as in *Lalage leucopyga*, the Long-tailed Triller

Leucorhoa *loy-ko-RO-a*
Greek, *leuko*, white, and *orrhos*, rump, as in *Oceanodroma leucorhoa*, Leach's Storm Petrel

Leucorodia *loy-kor-OH-dee-a*
Greek, *leuko*, white, and *rodo*, a rose, as in *Platalea leucorodia*, the Eurasian Spoonbill, which is almost all white, but takes on a rose wash if its food source contains plankton with red pigments

Loxia leucoptera,
Two-barred Crossbill

Leucosarcia *loy-ko-SAR-see-a*
Greek, *leuko*, white, and *sarcia*, a pack, bundle, or *sarc-*, flesh, as in *Leucosarcia melanoleuca*, the Wonga Pigeon, probably so named because people considered the bird a food source

Leucosticte *loy-ko-STICK-tee*
Greek, *leuko*, white, and *stictos*, varied, varicolored, as in *Leucosticte brandti*, Brandt's Mountain Finch

Leucotis *loy-KO-tis*
Greek, *leuko*, white, and *otos*, ear, as in *Basilinna leucotis*, the White-eared Hummingbird

Leucurus *loy-KOO-rus*
Greek, *leuko*, white, and *oura*, tail, as in *Elanus leucurus*, the White-tailed Kite

Levaillanti *le-va-LAN-tye*
After François Le Vaillant, French collector and naturalist, as in *Clamator levaillanti*, Levaillant's Cuckoo

Lewinii, -ia *loo-WIN-ee-eye/ee-a*
After John Lewin, English naturalist, as in *Meliphaga lewinii*, Lewin's Honeyeater

Lewis *LOO-wis*
After Meriwether Lewis, American explorer, as in *Melanerpes lewis*, Lewis's Woodpecker

Lichenostomis *lye-ken-o-STOME-is*
Greek, *leichen*, to lick, and *stoma*, mouth, as in *Lichenostomis* (now *Caligavis*) *chrysops*, the Yellow-faced Honeyeater

Lichmera *lik-MER-a*
Greek, *lichmeres*, flicking the tongue, as in *Lichmera limbata*, the Indonesian Honeyeater

Lichtensteinii *lik-ten-STINE-ee-eye*
After Martin Lichtenstein, German physician and ornithologist, as in *Pterocles lichtensteinii*, Lichtenstein's Sandgrouse

Limicola *li-mi-KO-la*
Limus, mud, and *cola*, dwell, as in *Rallus limicola*, the Virginia Rail

Limnocorax *lim-no-COR-aks*
Greek, *limne*, pond, marsh, lake, and *korax*, crow, raven, as in *Limnocorax* (now *Amaurornis*) *flavirostra*, the Black Crake

Limnoctites *lim-nok-TITE-eez*
Greek, *limne*, pond, marsh, lake, and *ktites*, dweller, as in *Limnoctites rectirostris*, the Straight-billed Reedhaunter

Limnodromus *lim-no-DRO-mus*
Greek, *limne*, pond, marsh, lake, and *dromeus*, runner, as in *Limnodromus griseus*, the Short-billed Dowitcher

Limnornis *lim-NOR-nis*
Greek, *limne*, pond, marsh, lake, and *ornis*, bird, as in *Limnornis curvirostris*, the Curve-billed Reedhaunter

Limnothlypis *lim-no-THLIP-is*
Greek, *limne*, pond, marsh, lake, and *thlypis*, small bird, as in *Limnothlypis swainsonii*, Swainson's Warbler

Limosa *li-MO-sa*
Limus, mud, and *osus*, full of, prone, as in *Limosa limosa*, the Black-tailed Godwit, the common name perhaps coming from Old English meaning good to eat

Lineatus, -a *lin-ee-AH-tus/a*
Striped or lined, as in *Buteo lineatus*, the Red-shouldered Hawk

Buteo lineatus,
Red-shouldered Hawk

Phalaropus lobatus,
Red-necked Phalarope

Liocichla *lye-o-SIK-la*
Greek, *lio*, smooth, and *cichla*, a thrush, as in *Liocichla ripponi*, the Scarlet-faced Liocichla, with smooth plumage of the back and face

Lioptilus *lye-op-TIL-us*
Greek, *lio*, smooth, soft, and *ptilion*, feather or wing, as in *Lioptilus nigricapillus*, the Bush Blackcap

Liosceles *ly-os-SEL-eez*
Greek, *lio*, smooth, soft, and *scelos*, leg, as in *Liosceles thoracicus*, the Rusty-belted Tapaculo, with few scales on legs, making them appear smooth

Littoralis *lit-to-RAL-is*
Shoreline, of the shore, as in *Ochthornis littoralis*, the Drab Water Tyrant, an inhabitant of river and streamsides

Livia *LIV-ee-a*
Livens, bluish, ashen, as in *Columba livia*, the Rock Dove

Lloydi *LOY-dye*
After William Lloyd, Irish-American collector, as in *Psaltriparus lloydi* (now *minimus*), the American Bushtit

Lobatus *lo-BA-tus*
Lobed toes, as in *Phalaropus lobatus*, the Red-necked Phalarope

Loboparadisea *lo-bo-par-a-DEES-ee-a*
Greek, *lobos*, lobe, and *paradise*, pleasure ground, as in *Loboparadisea sericea*, the Yellow-breasted Satinbird, with nasal lobes on its bill; once thought to be a Bird of Paradise

Lochmias *lock-MEE-as*
Greek, *lokhmaios*, inhabitant of the bush, as in *Lochmias nematura*, the Sharp-tailed Streamcreeper

Locustella *low-kus-TEL-la*
Locusta, locust, and *-ellus*, small, as in *Locustella fluviatilis*, the River Warbler, possibly named for its locust-like song

Loddigesia *lod-di-JEE-see-a*
After George Loddiges, British botanist and influential nurseryman, as in *Loddigesia mirabilis*, the Marvelous Spatuletail

Lomvia *LOM-vee-a*
Swedish for guillemot, murre, or diver, as in *Uria lomvia*, the Thick-billed Murre or Brunnich's Guillemot

Lonchura *lon-KOO-ra*
Greek, *lonkhe*, spearhead, and *oura*, tail, as in *Lonchura cucullata*, the Bronze Mannikin

Longicauda, -ta, -tus *lon-jee-KAW-da/lon-jee-kaw-DAT-a/us*
Longus, long, and *cauda*, tail, as in *Bartramia longicauda*, the Upland Sandpiper

Longipennis *lon-ji-PEN-nis*
Longus, long, and *penna*, feather, as in *Falco longipennis*, the Australian Hobby, with long wings

Longirostris *lon-ji-ROSS-tris*
Longus, long, and *rostris*, beak, as in *Rallus longirostris*, the Clapper Rail

Lophaetus *lo-FEE-tus*
Greek, *lophus*, crest, and *aetos*, eagle, as in *Lophaetus occipitalis*, the Long-crested Eagle

Rallus longirostris,
Clapper Rail

Lophodytes *lo-fo-DYE-teez*
Greek, *lophus*, crest, and *dytes*, diver, as in *Lophodytes cucullatus*, the Hooded Merganser

Lophoictinia *lo-fo-ik-TIN-ee-a*
Greek, *lophus*, crest, and *iktinos*, a kite, as in *Lophoictinia isura*, the Square-tailed Kite

Lopholaimus *lo-fo-LAY-mus*
Greek, *lophus*, crest, and *laimus*, throat, as in *Lopholaimus antarcticus*, the Topknot Pigeon

Lophonetta *lo-fo-NET-ta*
Greek, *lophus*, crest, and *netta*, duck, as in *Lophonetta specularioides*, the Crested Duck

Lophophorus *lo-fo-FOR-us*
Greek, *lophus*, crest, and *phorus*, bearer, as in *Lophophorus lhuysii*, the Chinese Monal

Lophortyx *lo-FOR-ticks*
Greek, *lophus*, crest, and *ortux*, quail as in *Lophortyx* (now *Callipepla*) *californica*, the California Quail

Lophostrix *lo-FO-stricks*
Greek, *lophus*, crest, and *strix*, owl, as in *Lophostrix cristata*, the Crested Owl

Lophotis, -tes *lo-FO-tis/teez*
Greek, *lophus*, crest, and *otis*, bustard, as in *Lophotis ruficrista*, the Red-crested Bustard / Korhaan

Lophozosterops *lo-fo-ZOS-ter-ops*
Greek, *lophus*, crest, *zoster*, a girdle, and *ops*, the eye, as in *Lophozosterops dohertyi*, the Crested White-eye

Lophura *lo-FOOR-a*
Greek, *lophus*, crest, and *oura*, tail, as in *Lophura swinhoii*, Swinhoe's Pheasant

Lorentzi *lo-RENTS-eye*
After Hendrik Lorentz, Dutch diplomat, as in *Pachycephala lorentzi*, Lorentz's Whistler

Loriculus *lor-ih-KOO-lus*
Malay, *lori*, parrot, and *culus*, small, as in *Loriculus galgulus*, the Blue-crowned Hanging Parrot; Hanging Parrots can sleep hanging upside down

Lorius *LOR-ee-us*
Malay, *lori*, parrot, as in *Lorius domicella*, the Purple-naped Lory

Loxia *LOCK-see-a*
Greek, *loxos*, crosswise, as in *Loxia scotica*, the Scottish Crossbill, the only vertebrate unique to the United Kingdom, with curved, overlapping mandibles to extract seeds from cones

Loxops *LOCKS-ops*
Greek, *loxos*, crosswise, and *ops*, the eye, as in *Loxops coccineus*, the Akepa; this refers to the slight crossing of the tips of the lower and upper jaws

---

LATIN IN ACTION

From the Middle French *bistarde*, meaning slow bird, we get bustard, as in the Red-crested Bustard, *Lophotis ruficrista*, also called the Korhaan. It is unique among bustards in having a retractile pink crest. Found from Eurasia to Australia, the 20 or so species of bustards are most common in Africa. They are large birds at 16 to 60 inches (40 to 150 centimeters) in length, with the Kori and Great Bustards often considered the world's heaviest flying birds, weighing up to 44 pounds (20 kilograms), although they rarely fly, sometimes going months without leaving the ground. They are adapted for a terrestrial lifestyle as they lack a hind toe and cannot perch on branches. All bustards are omnivorous and opportunistic and will feed on almost anything edible.

Lophotis ruficrista, Red-crested Bustard

Luciae *LOO-see-ee*
After Lucy Baird, daughter of Spencer Baird, as in *Leiothlypis luciae*, Lucy's Warbler

Lucidus *loo-SID-us*
Luci-, light, clear, shining, as in *Hemignathus lucidus*, the Nukupuu

Lucifer *LOO-si-fer*
Light bringing, as in *Calothorax lucifer*, the Lucifer Sheartail; probably refers to the bird's bright, iridescent violet throat

Ludlowi *LUD-lo-eye*
After Frank Ludlow, British educator, botanist, and ornithologist, as in *Fulvetta ludlowi*, the Brown-throated Fulvetta

Ludoviciana, -us *loo-doe-vee-see-AN-a/us*
Of Louisiana, as in *Piranga ludoviciana*, the Western Tanager

Lugubris *loo-GOO-bris*
Mourning, mournful, as in *Quiscalus lugubris*, the Carib Grackle; perhaps because the glossy black color reminds one of mourning

Lullula *lul-LOO-la*
From the bird's call, as in *Lullula arborea*, the Woodlark

Lunata, -us *loo-NA-ta/tus*
Lunatus, crescent-shaped, as in *Onychoprion lunatus*, the Spectacled Tern, probably alluding to its wing shape

Lunda *LOON-da*
Norwegian word for Puffin, as in *Lunda* (now *Fratercula*) *cirrhata*, the Tufted Puffin; Lundehunds are dogs bred especially to hunt puffins in their burrows

Luscinia *loo-SIN-ee-a*
Lusinius, a nightingale, as in *Luscinia calliope*, the Siberian Rubythroat

Lutea, -us *LOO-tee-a/us*
Luteus, yellow, as in *Leiothrix lutea*, the Red-billed Leiothrix or Pekin Nightingale, with yellow on the throat and breast

Luteifrons *LOO-tee-eye-fronz*
Luteus, yellow, and *frons*, forehead, as in *Nigrita luteifrons*, the Pale-fronted Nigrita

Luteiventris *loo-te-eye-VEN-tris*
Luteus, yellow, and *ventris*, underside, as in *Myiodynastes luteiventris*, the Sulphur-bellied Flycatcher

Piranga ludoviciana, Western Tanager

Lutosa *loo-TOW-sa*
Lutum, mud, as in *Caracara lutosa*, the extinct Guadalupe Caracara; possibly due to its predominantly dark brown plumage

Lybius *LIH-bee-us*
Probably a misspelling of Libya, as in *Lybius undatus*, the Banded Barbet

Lycocorax *ly-ko-KOR-aks*
Greek, *lyco*, wolf, and *corax*, raven, as in *Lycocorax pyrrhopterus*, the Paradise-crow

Lymnocryptes *lim-no-CRIP-teez*
Greek, *limne*, marsh, pond, and *kruptos*, hidden, as in *Lymnocryptes minimus*, the Jack Snipe, a bird rarely seen unless flushed

Lyrurus *lye-ROO-rus*
Greek, *lura*, a lyre, and *oura*, tail, as in *Lyrurus tetrix*, the Black Grouse

LOUIS AGASSIZ FUERTES
(1874–1927)

Louis Agassiz Fuertes was one of the most talented illustrators of birds in history. His detailed drawings and paintings continue to provide a storehouse of knowledge about bird species.

Born in Ithaca, New York, in 1874, Fuertes was named after Louis Agassiz, a renowned nineteenth-century Swiss naturalist. From early in his life Fuertes showed an extraordinary interest in birds. Although his father wanted his son, the youngest of six, to go into an established field, he realized how fascinated Louis was with birds when the eight year old captured an owl and tied it to the kitchen table. When Mr. Fuertes took his son to the Ithaca Public Library to show him Audubon's *Birds of America*, the boy found his vocation and began to draw birds in earnest.

Seeing how single-minded Louis became about killing and drawing birds, and afraid the boy would never be able to make a living as an artist, Fuertes's parents tried to shift his course by taking him to a preparatory school in Switzerland in 1892. The next year Fuertes unwillingly enrolled at Cornell as an architecture major. He failed almost all of his classes except drawing.

While still in college, he had an opportunity to show his bird illustrations to Elliott Coues, then one of the country's top ornithologists and a founding member of the American Ornithologists' Union. Coues was very impressed and convinced him that he could support himself as an artist. Taking the twenty year old under his wing, Coues introduced Louis to the field of ornithology, displayed his artwork at an American Ornithologists' Union meeting, and encouraged him to obtain commissions for his drawings.

Fuertes did more than 100 drawings for Mabel Osgood Wright and Eliot Coues's *Citizen Bird: Scenes from Bird-Life in Plain English for Beginners*, between 1896 and 1897. In 1899, Edward Harriman, a wealthy railroad magnate, arranged a scientific exploration of the coast of Alaska and brought with him several respected scientists such as C. Hart Merriam, John Muir, Robert Ridgway, and others. He included two photographers and three artists, including Fuertes. Fuertes often killed and skinned birds to study them closely, but when he could not, he made quick sketches and described their songs in his notes.

When Fuentes's detailed, full-color drawings from the Alaska expedition were published, he became well known and in demand for his works. Some of the works he illustrated include *Handbook of Birds of Western North America*

Neophron percnopterus,
Egyptian Vulture

Fuertes' illustration of an Egyptian Vulture. Although vultures feed mainly on carrion, only New World vultures can detect their potential meal by smell.

by Frank Chapman, 1902; *Upland Game Birds* by Edwyn Sandys and T. S. van Dyke, 1902; *Key to North American Birds* by Elliot Coues, 1903; and *Birds of New York* by Elon Howard Eaton, 1910.

Although Fuertes graduated with a degree in architecture, he became a lecturer in ornithology at Cornell, but took a leave of absence from his lecturing position to accompany Dr. Wilfred Osgood of Chicago's Field Museum of Natural History to Abyssinia (now Ethiopia). He produced some of his finest sketches, rendering lifelike images of birds from brief glimpses. His phenomenal memory enabled him to reproduce an individual bird in complete detail even years after being in the field.

Fuertes died in a car accident with a train in 1927, shortly after returning from Abyssinia. In his lifetime he had prepared 3,500 bird skins and drawn over 1,000 studio and field sketches of over 400 species of birds from all over the world.

Fuertes is considered to have produced the most realistic bird illustrations of his time and they are still in demand. A recent auction of his Wild Turkey sold for over $86,000 in 2012.

Accipiter striatus, Sharp-shinned Hawk

As Fuertes's illustration shows, the Sharp-shinned Hawk and the larger Cooper's Hawk are virtually identical twins except for size.

Whistling Ducks are not true ducks but in a separate subfamily. On the right is the Fulvous Whistling Duck, *Dendrocygna bicolor*. It is unclear what species Fuertes intended the other two whistling ducks to be.

"If the birds of the world had met to select a human being who could best express to mankind the beauty and charm of their forms… they would unquestionably have chosen Louis Fuertes."

Dr. Chapman speaking at Fuertes' funeral, quoted in American National Biography

M

Macgillivrayi mak-GIL-li-vray-eye
After John MacGillivray, Australian naturalist, and son of William MacGillivray, Scottish artist and professor. The Fiji Petrel, *Pseudobulweria macgillivrayi*, is named after the son and MacGillivray's Warbler, *Geothlypis tolmiei*, after the father

Machetornis mak-eh-TOR-nis
Greek, *makhetes*, fighter, and *ornis*, bird, as in *Machetornis rixosa*, the Cattle Tyrant

Mackinlayi mak-KIN-lee-eye
After Archibald Mackinlay, as in *Macropygia mackinlayi*, Macinlay's Cuckoo-Dove

Macrocephalon mak-ro-se-FAL-on
Greek, *macro*, long or large, and Latin, *cephala*, head, as in *Macrocephalon maleo*, the Maleo

Macrodactyla mak-ro-dak-TIL-a
Greek, *macro*, long or large, and *dactylos*, finger, toe, as in *Oceanodroma macrodactyla*, the probably extinct Guadalupe Storm Petrel, whose middle toe and claw are twice as long as its nearest relative's

Macrodipteryx mak-ro-DIP-ters-iks
Greek, *macro*, long or large, *di-*, two, and *pteryx*, wing, as in *Macrodipteryx longipennis*, the Standard-winged Nightjar

Macronectes mak-ro-NEK-teez
Greek, *macro*, long or large, and *nekes*, a swimmer, as in *Macronectes giganteus*, the Southern Giant Petrel

Macronyx mak-RON-iks
Greek, *macro*, long or large, and *onux*, claw, as in *Macronyx ameliae*, the Rosy-breasted, or Rosy-throated Longclaw

Macrorhynchus mak-ro-RINK-us
Greek, *macro*, long or large, and Latin, *rhynchus*, bill, as in *Saxicola macrorhynchus*, the White-browed Bush Chat

Macroura, -us mak-ROO-ra/rus
Greek, *macro*, long or large, and *oura*, tail, as in *Zenaida macroura*, the Mourning Dove, named for its plaintive call; also as in *Vidua macroura*, the Pin-tailed Whydah

Macularia, -us mak-oo-LAR-ee-a/us
Macula, spot, as in *Actitis macularius*, the Spotted Sandpiper

Maculata, -um, -us mak-oo-LAT-a/um/us
Macula, spot, as in *Stachyris maculata*, the Chestnut-rumped Babbler, with a heavily spotted chest and belly

Maculicauda, -us mak-oo-li-KAW-da/dus
Macula, spot, and *cauda*, tail, as in *Hypocnemoides maculicauda*, the Band-tailed Antbird, with spots on its tail

Maculicoronatus mak-oo-li-cor-o-NAT-us
Macula, spot, and *corona*, crown, as in *Capito maculicoronatus*, the Spot-crowned Barbet

Maculifrons mak-OO-li-fronz
Macula, spot, and *frons*, forehead, brow, as in *Veniliornis maculifrons*, the Yellow-eared Woodpecker, with spots on its forehead

Maculipectus mak-oo-li-PEK-tus
Macula, spot, and *pectus*, breast, as in *Pheugopedius maculipectus*, the Spot-breasted Wren

Maculipennis mak-oo-li-PEN-nis
Macula, spot, and *penna*, feather, as in *Chroicocephalus maculipennis*, the Brown-hooded Gull

Zenaida macroura, Mourning Dove

Maculirostris *mak-oo-li-ROSS-tris*
Macula, spot, and *rostris*, bill, as in *Muscisaxicola maculirostris*, the Spot-billed Ground Tyrant

Maculosa, -us *mak-oo-LO-sa/sus*
Spotted, as in *Nothura maculosa*, the Spotted Nothura

Madagascariensis, -inus
mad-a-gas-kar-ee-EN-sus/EYE-nus
After Madagascar, as in *Caprimulgus madagascariensis*, the Madagascan Nightjar

Magellanica, -us *ma-jel-LAN-ih-ka/kus*
Straits of Magellan, as in *Spinus magellanica*, the Hooded Siskin

Magna, -num *MAG-na/num*
Large, as in *Sturnella magna*, the Eastern Meadowlark; may refer to either the bird's range or its size compared with starlings

Magnificens, -cus *mag-NIF-ih-senz/kus*
Splendid, grand, as in *Fregata magnificens*, the Magnificent Frigatebird

Magnirostris, -tre, -tra
mag-ni-ROSS-tris/tree/tra
Magna, large, and *rostris*, beak, as in *Gerygone magnirostris*, the Large-billed Gerygone

Magnolia *mag-NO-lee-a*
Latinized name of Pierre Magnol, French physician and botanist, as in *Setophaga magnolia*, the Magnolia Warbler

Major *MAY-jor*
Maior, large, great, as in *Locustella major*, the Long-billed Bush Warbler

Malabaricus, -ka *mal-a-BAR-ih-kus/ka*
After Malabar, a region of India, as in *Copsychus malabaricus*, the White-rumped Shama

Malacca, -ensis *mal-AK-ka/mal-a-KEN-sis*
After Malacca, Malaysia, as in *Lonchura malacca*, the Tricolored Munia

Malacocincla *mal-a-ko-SINK-la*
Greek, *malakos*, soft, and *cincla*, thrush, as in *Malacocincla cinereiceps*, the Ashy-headed Babbler; resembles a thrush and has soft-appearing plumage

Setophaga magnolia,
Magnolia Warbler

Malaconotus *mal-a-kon-O-tus*
Greek, *malakos*, soft, and *noton*, back, south end, as in *Malaconotus cruentus*, the Fiery-breasted Bushshrike, with soft-appearing feathers on its back

Malacopteron *mal-a-KOP-ter-on*
Greek, *malakos*, soft, and *pteron*, wing, as in *Malacopteron albogulare*, the Gray-breasted Babbler

Malacoptila *mal-a-cop-TIL-a*
Greek, *malakos*, soft, and *ptila*, feather, as in *Malacoptila panamensis*, the White-whiskered Puffbird

Malacorhynchus *mal-a-ko-RINK-us*
Greek, *malakos*, soft, Latin, *rhynchus*, bill, as in *Malacorhynchus membranaceus*, the Pink-eared Duck; the large spoon-shaped bill appears soft and pliable

Maleo *MAL-ee-o*
From the native Indonesian name, as in *Macrocephalon maleo*, the Maleo

Malherbi *mal-ERB-ee-eye*
After Alfred Malherbe, French magistrate and naturalist, as in *Cyanoramphus malherbi*, Malherbe's Parakeet

Malurus *mal-OO-rus*
Greek, *malos*, soft, and *oura*, tail, as in *Malurus cyaneus*, the Superb Fairywren

Manacus *man-AH-kus*
Latin, from the Dutch *manneken*, a dwarf, little man, as in *Manacus manacus*, the White-bearded Manakin

Manucodia *man-oo-KO-dee-a*
Javanese, *manuk dewata*, bird of the gods, as in *Manucodia comrii*, the Curl-crested Manucode

Mareca *mar-EK-a*
From the Portuguese for a kind of duck, as in *Mareca* (now *Anas*) *penelope*, the Eurasian Wigeon

Margaritae *mar-gar-EE-tee*
After Margaret Holt, the wife of American ornithologist E. G. Holt, as in *Batis margaritae*, Margaret's Batis

Margaroperdix *mar-gar-o-PER-diks*
Greek, *margarodes*, pearlescent, and *perdix*, partridge, as in *Margaroperdix madagarensis*, the Madagascan Partridge

Margarops *MAR-ga-rops*
Greek, *margarites*, pearl, and *opsis*, appearance, look, as in *Margarops fuscatus*, the Pearly-eyed Thrasher

Margarornis *mar-gar-OR-nis*
Greek, *margarodes*, pearlescent, and *ornis*, bird, as in *Margarornis squamiger*, the Pearled Treerunner

Marginata, -us *mar-jin-AT-a/us*
Marginatus, rimmed, as in *Charadrius marginatus*, the White-fronted Plover

Marila *mar-IL-a*
Greek, *marile*, charcoal, as in *Aythya marila*, the Greater Scaup, with overall dark coloration

Marina *mar-EE-na*
Marine, of the sea, as in *Pelagodroma marina*, the White-faced Storm Petrel

Marinus *mar-EE-nus*
Marine, of the sea, as in *Larus marinus*, the Great Black-backed Gull

Maritima, -mus *mar-ih-TEE-ma/mus*
Marine, of the sea, as in *Ammodramus maritimus*, the Seaside Sparrow

Markhami *MARK-am-eye*
After Albert Markham, English explorer and navy admiral, as in *Oceanodroma markhami*, the Markham's Storm Petrel

Marmoratus *mar-mo-RA-tus*
Marbled, of marble, as in *Brachyramphus marmoratus*, the Marbled Murrelet

Martinica, -us *mar-tin-EE-ka/kus*
After Martinique, a Caribbean island, as in *Porphyrio martinicus*, the (American) Purple Gallinule, gallinule derived from the Latin *gallina*, meaning little hen

Mauri *MAW-rye*
After Ernesto Mauri, Italian botanist, as in *Calidris mauri*, the Western Sandpiper

Maximiliani *maks-i-mil-ee-AN-eye*
After Prince Philipp Maximilian, German aristocrat and explorer, as in *Pionus maximiliani*, the Scaly-headed Parrot

Maximus, -a *MAKS-ee-mus/ma*
Largest, greatest, as in *Thalasseus maximus*, the Royal Tern

Thalasseus maximus, Royal Tern

Mayri *MARE-eye*
After Ernst Mayr, German ornithologist and evolutionary biologist, as in *Ptiloprora mayri*, Mayr's Honeyeater

Mayrornis *mare-OR-nis*
After Ernst Mayr, German evolutionary biologist and ornithologist, and Greek *ornis*, bird, as in *Mayrornis versicolor*, the Versicolored Monarch

Mccownii *mak-KOWN-ee-eye*
After John McCown, American Army officer and naturalist, as in *Rhynchophanes mccownii*, McCown's Longspur

Meeki *MEEK-eye*
After Albert Meek, English explorer, as in *Ninox meeki*, the Manus Boobook or Hawk-Owl

Megaceryle *me-ga-sir-IL-ee*
Greek, *mega*, great, large, and *ceryle*, a kingfisher, as in *Megaceryle alcyon*, the Belted Kingfisher

Megadyptes *me-ga-DIP-teez*
Greek, *mega*, great, large, and *dyptes*, diver, as in *Megadyptes antipodes*, the Yellow-eyed Penguin

Megalaima *me-ga-LAY-ma*
Greek, *mega*, great, large, and *laima*, throat, as in *Megalaima chrysopogon*, the Golden-whiskered Barbet (see box)

Megalopterus *me-ga-LOP-ter-us*
Greek, *mega*, great, large, and *ptery*, winged, as in *Phalcoboenus megalopterus*, the Mountain Caracara

Megapodius *me-ga-POD-ee-us*
Greek, *mega*, great, large, and *pous*, foot, as in *Megapodius laperouse*, the Micronesian Megapode

Megarynchus, -a, -os *me-ga-RINK-us/a/os*
Greek, *mega*, great, large, and Latin, *rhynchus*, bill, as in *Megarynchus pitangua*, the Boat-billed Flycatcher

Megascops *MEG-a-skops*
Greek, *mega*, great, large, and *scops*, owl, as in *Megascops nudipes*, the Puerto Rican Screech Owl

Melaenornis *mel-ee-NOR-nis*
Greek, *melas*, black, dark, and *ornis*, bird, as in *Melaenornis pammelaina*, the Southern Black Flycatcher

---

LATIN IN ACTION

The faces of barbets have stiff bristles that extend forward and cover the nares, the base of the jaw, and the neck region, thus resembling a beard; hence the name barbet from the Latin *barbatus*, bearded. The Golden-whiskered Barbet, along with 25 other species, belongs to the family of Asian barbets, the Megalaimidae, which reflects their characteristic large throats. Megalaimidae have zygodactylous feet and are brightly colored green with red, blue, and yellow markings. *Megalaima chrysopogon* describes the Golden-whiskered Barbet as large-throated with a gold (*chryso*) beard (Greek, *pogon*).

Megalaima chrysopogon, Golden-whiskered Barbet

Melancholicus *mel-an-KOL-ih-kus*
Greek, *melas*, black, dark, and *chol-e*, bile, as in *Tyrannus melancholicus*, the Tropical Kingbird, an aggressive rather than melancholic bird

Melanerpes *mel-an-ER-peez*
Greek, *melas*, black, dark, and *herpes*, creeper, as in *Melanerpes formicivorus*, the Acorn Woodpecker

Melanitta *mel-an-NIT-ta*
Greek, *melas*, black, dark, and *netta*, duck, as in *Melanitta fusca*, the Velvet Scoter

MELANOCEPHALA

Melanocephala, -us *mel-an-o-se-FAL-a/us*
Greek, *melas*, black, dark, and Latin, *cephala*, head, as in *Arenaria melanocephala*, the Black Turnstone

Melanoceps *mel-AN-o-seps*
Greek, *melas*, black, dark, and Latin *ceps*, headed, as in *Myrmeciza melanoceps*, the White-shouldered Antbird

Melanochlamys *mel-an-o-KLAM-is*
Greek, *melas*, black, dark, and *chlamy*, cloak, as in *Accipiter melanochlamys*, the Black-mantled Goshawk

Melanochlora *mel-an-o-KLOR-a*
Greek, *melas*, black, dark, and *khloros*, green, as in *Melanochlora sultanea*, the Sultan Tit, with a glossy blackish-green back, neck, and throat

Melanocorypha *mel-an-o-kor-IF-a*
Greek, *melas*, black, dark, and *koryphe*, head, as in *Melanocorypha mongolica*, the Mongolian Lark

Melanocorys *mel-an-o-KOR-is*
Greek, *melas*, black, dark, and *korus*, lark, as in *Calamospiza melanocorys*, the Lark Bunting

Melanogaster *mel-an-o-GAS-ter*
Greek, *melas*, black, dark, and *gastro-*, stomach, as in *Ploceus melanogaster*, the Black-billed Weaver

Melanogenys *mel-an-o-JEN-is*
Greek, *melas*, black, dark, and *genys*, cheek, as in *Adelomyia melanogenys*, the Speckled Hummingbird

Melanoleuca, -os, -us *mel-an-o-LOY-kak/os/kus*
Greek, *melas*, black, dark, and *leukos*, white, as in *Tringa melanoleuca*, the Greater Yellowlegs

Melanolophus *mel-an-o-LO-fus*
Greek, *melas*, black, dark, and *lophus*, crest, as in *Gorsachius melanolophus*, the Malayan Night Heron

Melanonota, -us *mel-an-o-NO-ta/us*
Greek, *melas*, black, dark, and *nota*, mark, as in *Pipraeidea melanonota*, the Fawn-breasted Tanager

Melanophris *mel-an-O-friss*
Greek, *melas*, black, dark, and *ophris*, eybrow, as in *Thalassarche melanophris*, the Black-browed Albatross

Melanops *MEL-an-ops*
Greek, *melas*, black, dark, and *ops*, eye, as in *Centropus melanops*, the Black-faced Coucal

Melanoptera, -us *mel-an-OP-ter-a/us*
Greek, *melas*, black, dark, and *pteron*, wing, as in *Coracina melanoptera*, the Black-headed Cuckooshrike

Melanospiza *mel-an-o-SPY-za*
Greek, *melas*, black, dark, and *spiza*, finch, as in *Melanospiza richardsoni*, the St. Lucia Black Finch

Melanotis *mel-an-O-tis*
Greek, *melas*, black, dark, and *otus*, ear, as in *Pteruthius melanotis*, the Black-eared Shrike-babbler

Melanotos *mel-an-O-tos*
Greek, *melas*, black, dark, and *noton*, the back, as in *Calidris melanotos*, the Pectoral Sandpiper

Melanura, -us *mel-an-OO-ra/us*
Greek, *melas*, black, dark, and *oura*, tail, as in *Polioptila melanura*, the Black-tailed Gnatcatcher

Thalassarche melanophris, Black-browed Albatross

MELANERPES

From the Greek, *melas*, black, dark, and *herpes*, creeper, the genus *Melanerpes* (mel-an-ER-peez) is the largest of the 30 genera of woodpeckers with 22 species out of a total of 200. The only places woodpeckers are not found are Australia, New Zealand, Madagascar, and the polar regions. They are all very recognizable with their stiff tails, zygodactyl feet (two toes forward, two back), and their habit of climbing trees vertically and pecking at the bark.

The specific names of the *Melanerpes* genus tend to be nicely descriptive. There is *M. aurifrons*, the Golden-fronted Woodpecker; *M. formicivorous*, the ant-eating Acorn Woodpecker; and *M. erythrocephalus*, the Red-headed Woodpecker. There are also several eponyms like *M. hoffmannii*, *M. lewis*, and *M. pucherani*.

The most fascinating aspect of all woodpeckers is how they peck at bark and drill holes in trees or fences. Woodpeckers hammer their beaks into trees 18 to 22 times per second, at speeds of 13 to 15 miles per hour (21 to 25 kilometers per hour), thousands of times per day, subjecting their brains to deceleration forces of 1.2 kilograms with each strike. So what is it about the woodpecker skull that protects them?

Melanerpes aurifrons, Golden-fronted Woodpecker

The beak is hard but elastic; the lower bill bends slightly with each impact. The skull is constructed of a large number of thin bones that criss-cross each other, making the head spongy and able to deform a little. A special bone called the hyoid supports the tongue and wraps around the back of the skull to the nasal openings. Covered with muscles, the hyoid bone's looping structure around the whole skull acts like a safety belt.

The musculature of the tongue allows it to be extended the length of the head or more. The tongue is made sticky by secretions from salivary glands and the tip of the tongue, furnished with barbs or spines, can be manipulated to capture insects or larvae.

Besides displaying these amazing adaptatione, woodpeckers serve a very important function in the environment by providing holes for other birds to nest in. Bluebirds, tits, nuthatches, wrens, and others rely on woodpecker-provided cavities.

Melanerpes herminieri, Guadeloupe Woodpecker

The Guadeloupe Woodpecker is endemic to the island of Guadeloupe. In spite of habitat degradation, its population is stable.

Meleagris

This genus consists of two species: the Wild Turkey of discontinuous distribution across the United States and the Oscellated Turkey, found only in the Yucatan region of Central America. *Meleagris* (*mel-ee-AH-gris*) comes from the Latin meaning guinea fowl; the Wild Turkey specific name is *gallopavo* (Latin *gallo*, cock, and *pavo*, peacock), and that of the Ocellated Turkey is *ocellata*. No one knows where the word turkey came from, but it might have been Columbus who called it tuka or tukki.

We also don't know if pilgrims and Indians ate Thanksgiving turkey in seventeen-century North America, because apparently the pilgrims called all wildfowl turkey. The Spaniards brought the Wild Turkey home from North America and it gradually became popular all over Europe, being variously called turkey-fowl, turkey bird, turkey cock, and even Indian Fowl because it was thought to come from the West Indies. When the

Meleagris gallopavo, Wild Turkey

colonists were introduced to the bird by Native Americans, they were surprised to see a bird they were familiar with since it had been raised in England for many generations by this time.

As American pioneers moved westward and cleared the forests, the habitat for turkeys diminished. By the mid-1800s the turkey was gone from almost half of its original range, and by the early 1900s only around 30,000 turkeys remained in the wild. After the turn of the century the decline of the turkey population halted as protective measures and reintroductions brought the population up to about 4.5 million across all states except Alaska.

Turkeys are now bred to have a higher protein level and bigger breast muscles than wild birds, and are raised in open pens or environmentally controlled barns. The population of the US eats nearly 300 million turkeys each year, and around 50 million of those are consumed at Thanksgiving. That is about 8 kilograms per year per person. In the EU, about 3.5 kilograms are eaten per person each year. Australians and South Africans eat a mere kilogram of turkey each year, mainly around Christmas.

Benjamin Franklin wanted the turkey to be the national symbol of the US. The Bald Eagle, *Haliaeetus leucocephalus*, won, but the turkey graces tables nicely.

Meleagris ocellata, Ocellated Turkey

Meleagris *mel-ee-AH-gris*
Greek, guineafowl; early dictionaries interchanged words for turkey, guineafowl, and peafowl, as in *Meleagris gallopavo*, the Wild Turkey, and *Numida meleagris*, the Helmeted Guineafowl

Melichneutes *mel-ik-NOY-teez*
Greek, *meli*, honey, and *ikhnos*, a track, footstep, as in *Melichneutes robustus*, the Lyre-tailed Honeyguide

Melidectes *mel-ee-DEK-teez*
Greek, *meli*, honey, and *dektes*, a beggar, as in *Melidectes leucostephes*, the Vogelkop Melidectes of the honeyeater family

Melierax *mel-ee-AIR-aks*
Greek, *melos*, song, and *hierax*, hawk or falcon, as in *Melierax canorus*, the Pale Chanting Goshawk

Melilestes *mel-ee-LES-teez*
Greek, *meli*, honey, and *lestes*, a thief, as in *Melilestes megarhynchus*, the Long-billed Honeyeater

Meliphaga *mel-ee-FA-ga*
Greek, *meli*, honey, and *phagein*, eat, as in *Meliphaga gracilis*, the Graceful Honeyeater

Melithreptus *mel-ee-THREP-tus*
Greek, *meli*, honey, and *threptos*, feed, nourish, as in *Melithreptus albogularis*, the White-throated Honeyeater

Mellisuga *mel-li-SOO-ga*
Mel, honey, and *sugo*, suck, as in *Mellisuga helenae*, the Bee Hummingbird

Melodia *mel-O-dee-a*
Greek, *melodos*, melodious, as in *Melospiza melodia*, the Song Sparrow

Melodus *mel-O-dus*
Greek, *melodos*, melodious, as in *Charadrius melodus*, the Piping Plover

Melophus *mel-O-fus*
Greek, *melas*, black, dark, and *lophus*, crest, as in *Melophus* (now *Emberiza*) *lathami*, the Crested Bunting

Melopsittacus *mel-op-SIT-ta-kus*
Greek, *melos*, song, and Latin, *psittacus*, parrot, as in *Melopsittacus undulatus*, the Budgerigar or Common Parakeet

Melopyrrha *mel-o-PEER-a*
Greek, *melas*, black, dark, and *pyrrha*, red, flame-colored, as in *Melopyrrha nigra*, the Cuban Bullfinch

Melospiza *mel-o-SPY-za*
Greek, *melos*, song, and *spiza*, finch, as in *Melospiza lincolnii*, Lincoln's Sparrow

Membranaceus *mem-bra-NAY-see-us*
Membrana, membranous, as in *Malacorhynchus membranaceus*, the Pink-eared Duck; the scientific name describes the pliable spoon-shaped bill with membranes for filter feeding

Menckei *MENK-ee-eye*
After Bruno Mencke, German zoologist, as in *Symposiachrus menckei*, the Mussau Monarch, of Mussau Island, New Guinea

Mentalis *men-TAL-is*
Of the chin, as in *Cracticus mentalis*, the Black-backed Butcherbird. This bird has a white chin set off from its mostly black head and neck

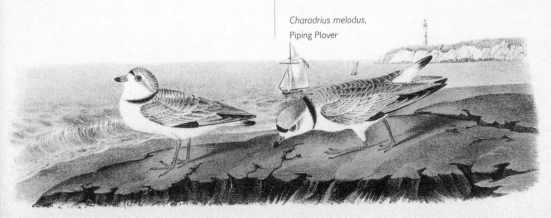

Charadrius melodus, Piping Plover

Menura men-OO-ra
Greek, *mene*, moon, and *oura*, tail, as in *Menura alberti*, Albert's Lyrebird, with crescent moon-like markings (lunules) on the inner web of the outer tail feathers

Merganetta mer-gan-ET-ta
Mergus, diver, and *netta*, duck, as in *Merganetta armata*, the Torrent Duck

Merganser mer-GAN-zer
Merger, to dive, and *anser*, goose, as in *Mergus merganser*, the Common Merganser

Meropogon mer-o-PO-gon
Greek, *merops*, a bee-eater, and *pogon*, beard, as in *Meropogon forsteni*, the Purple-bearded or Celebes Bee-eater

Merops MER-ops
Greek, *merops*, a bee-eater, as in *Merops apiaster*, the European Bee-eater

Merrilli MER-ril-eye
After Elmer Merrill, American botanist, as in *Ptilinopus merrilli*, the Cream-breasted Fruit Dove

Mexicana, -us, -um mecks-ih-KAN-a/us/um
After Mexico, as in *Sialia mexicana*, the Western Bluebird

Meyeri, -ianus MY-er-eye/my-er-ee-AN-nus
After Adolf Meyer, German anthropologist and ornithologist, as in *Epimachus meyeri*, the Brown Sicklebill

Micrastur my-KRAS-ter
Greek, *micros*, small, and Latin, *astur*, hawk, as in *Micrastur ruficollis*, the Barred Forest Falcon

Microcochlearius my-kro-ko-klee-AR-ee-us
Greek, *micros*, small, and Latin, *cochlear*, spoon, as in *Microcochlearius* (now *Hemitriccus*) *josephinae*, the Boat-billed Tody-Tyrant

Microhierax my-kro-HY-er-aks
Greek, *micros*, small, and *hierax*, a hawk or falcon, as in *Microhierax melanoleucos*, the Pied Falconet

Micromegas my-kro-MAY-gas
Greek, *micros*, small, and *mega*, large, as in *Nesoctites micromegas*, the Antillean Piculet, which is twice as large as any of the other piculets (a subfamily of small woodpeckers)

Micromonacha my-kro-mo-NAK-a
Greek, *micros*, small, and *monakhos*, a monk, as in *Micromonacha lanceolata*, the Lanceolated Monklet; to someone this bird with a lance-shaped bill resembled a monk

Micropalama my-kro-pa-LAM-a
Greek, *micros*, small, and *palama*, palm, as in *Micropalama* (now *Calidris*) *himantopus*, the Stilt Sandpiper, referring to the small amount of webbing between the toes

Micropsitta my-krop-SIT-ta
Greek, *micros*, small, and Latin *psittacus*, parrot, as in *Micropsitta geelvinkiana*, the Geelvink Pygmy Parrot

Microptera, -us my-KROP-ter-a/us
Greek, *micros*, small, and *pteron*, wing, as in *Mirafra microptera*, the Burmese Bush Lark

Microrhynchus, -um my-kro-RINK-us/um
Greek, *micros*, small, and Latin, *rhynchus*, bill, as in *Bradornis microrhynchus*, the African Gray Flycatcher

Microsoma my-kro-SO-ma
Greek, *micros*, small, and *soma*, body, as in *Oceanodroma microsoma*, the Least Storm Petrel

Sialia mexicana, Western Bluebird

Migratorius my-gra-TOR-ee-us
Migrare, to move, as in *Turdus migratorius*, the American Robin, migratory in parts of its range

Militaris mil-ih-TAR-is
Militar-, soldier, as in *Ara militaris*, the Military Macaw

Milvago mil-VA-go
Milvus, bird of prey, and *-ago*, resembling, as in *Milvago chimango*, the Chimango Caracara

Milvus MIL-vus
Milvus, bird of prey, as in *Milvus migrans*, the Black Kite

Mimus MIM-us
Mimic, imitator, as in *Mimus polyglottos*, the Northern Mockingbird, which copies the songs of other bird species

Mindanensis min-da-NEN-sis
After Mindinao, Phillipines, as in *Coracina mindanensis*, the Black-bibbed Cicadabird

Minimus, -um, -a MIN-ih-mus/mum/ma
Least, or smallest, as in *Psaltriparus minimus*, the American Bushtit

Mino MY-no
Mino, from the Hindi, *maina*, mynah, as in *Mino* (now *Ampeliceps*) *coronatus*, the Golden-crested Myna

Minor MY-nor
Inferior in grade, age, as in *Chordeiles minor*, the Common Nighthawk. It is a medium-sized nighthawk, but probably seemed small at the time it was named (when smaller species were unknown)

Minutilla myn-oo-TIL-la
Minutus, little, as in *Calidris minutilla*, the Least Sandpiper

Minutus, -a my-NOO-tus/a
Very small, as in *Hydrocoloeus minutus*, the Little Gull

Mirabilis mir-AH-bi-lis
Wonderful, as in *Eriocnemis mirabilis*, the Colorful Puffleg

Mirafra mir-AF-ra
Miras, wonderful, and *afra*, African, as in *Mirafra africana*, the Rufous-naped Lark (see box)

Mississippiensis mis-si-sip-pee-EN-sis
After Mississippi, as in *Ictinia mississippiensis*, the Mississippi Kite

LATIN IN ACTION

From the Old English *laferce*, meaning songbird, comes the word lark. The melodious songs of many larks gave rise to the saying "happy as a lark" and the rather carefree attitude of "going off on a lark." Almost all larks are found in the Old World and parts of Australia, and only one, the Horned or Shore Lark, *Eremophila alpestris*, is found in North America. (The so-called meadowlarks are actually classified as blackbirds.) The Rufous-naped Lark's genus *Mirafra* (from *mira*, wonderful) probably refers to its call. Larks are ground dwellers (with an extended rear toe), but to advertise their territories and attract mates, they have developed complex "flight songs" that they often deliver in mid-air.

Mirafra africana, Rufous-naped Lark

Mniotilta nee-o-TIL-ta
Greek, *mnion*, moss, and *tiltos*, plucked, as in *Mniotilta varia*, the Black-and-white Warbler, which uses moss and other items to construct its nest

Modesta, -tus mo-DES-ta/tus
Modestus, restrained, mild, modest, as in *Progne modesta*, the Galapagos Martin, with plain, unadorned plumage

Mollissima mol-LISS-sim-a
Mollis, soft, as in *Somateria mollissima*, the Common Eider, a bird whose down is collected from its nests to be used for pillows and such

Monarcha godeffroyi,
Yap Monarch

Molluccensis *mol-luk-SEN-sis*
After the Moluccas (Maluku Islands), as in *Pitta moluccensis,* the Blue-winged Pitta

Molothrus *mol-O-thrus*
Greek, *molobrus,* beggar or parasite, as in *Molothrus ater,* the Brown-headed Cowbird, which is a nest parasite

Momotus *mo-MO-tus*
Derives from the bird's call, motmot, as in *Momotus momota,* the Amazonian Motmot

Monachus *mo-NAK-us*
Monk, as in *Myiopsitta monachus,* the Monk Parakeet; the hood-like markings on the head are said to resemble a monk's hood

Monarcha *mo-NAR-ka*
Greek, *monarkhos,* monarch, king, as in *Monarcha godeffroyi,* the Yap Monarch

Monasa *mo-NAS-a*
Greek, *monases,* alone, as in *Monasa flavirostris,* the Yellow-billed Nunbird, a sedentary bird. They live in small, territorial groups and keep to themselves

Mongolica *mon-GO-lik-a*
After Mongolia, as in *Melanocorypha mongolica,* the Mongolian Lark

Monias *mo-NYE-as*
Greek, *monases,* alone, as in *Monias benschi,* the Subdesert Mesite; the scientific name a misnomer, as it is found in groups

Monocerata *mon-o-ser-AH-ta*
Greek, *monos,* single, one, and *keras,* horn, as in *Cerorhinca monocerata,* the Rhinoceros Auklet

Montana, -us *mon-TAN-a/us*
Relating to mountains, as in *Charadrius montanus,* the Mountain Plover

Montani *mon-TAN-eye*
After Joseph Montano, French anthropologist, as in *Anthracoceros montani,* the Sulu Hornbill

Montezumae *mon-te-ZOOM-ee*
Latinized form of the name of the Aztec emperor of Mexico, as in *Cyrtonyx montezumae,* the Montezuma Quail

Monticola *mon-ti-KO-la*
Montis, mountain, and *colo,* inhabit, as in *Monticola brevipes,* the Short-toed Rock Thrush

Montifringilla *mon-ti-frin-JIL-la*
Montis, mountain, and *fringilla,* finch, as in *Fringilla montifringilla,* the Brambling

Morinellus *mor-ih-NEL-lus*
Greek, *moros,* foolish, stupid, and *ella,* diminutive, as in *Charadrius morinellus,* the Eurasian Dotterel, a bird easily approached

Morus *MOR-us*
Greek, *moros,* foolish, stupid, as in *Morus bassanus,* the Northern Gannet, possibly named for its spectacular feeding dives

Motacilla *mo-ta-SIL-la*
Motus, move, and *cilla,* inaccurately used to mean tail, as in *Motacilla alba,* the White Wagtail, which frequently wags its tail

Muelleri *MEW-ler-eye*
After Salomon Müeller, a Dutch naturalist, as in *Lewinia muelleri,* the Aukland Rail

Multistriata, -us *mul-ti-stree-AT-a/us*
Multi, many, and *striata,* a streak, furrow, as in *Charmosyna multistriata,* the Striated Lorikeet

Muscicapa *mus-si-KAP-a*
Musca, fly, and *capio,* capture, as in *Muscicapa* (now *Cyanoptila*) *cyanomelana,* the Blue and White Flycatcher

Muscisaxicola *mus-si-saks-ih-KO-la*
Musca, fly, *saxum*, stone, and *colo*, inhabit, as in *Muscisaxicola maculirostris*, the Spot-billed Ground Tyrant, a flycatcher that nests on the ground

Muscivora *mus-si-VOR-a*
Musca, fly, and *vorus*, devour, swallow, as in *Muscivora* (now *Tyrannus*) *forficatus*, the Scissor-tailed Flycatcher

Musophaga *moo-so-FAY-ga*
Musa, banana, and *phagus*, eater of, as in *Musophaga rossae*, Ross's Turaco

Mustelina *mus-tel-EE-a*
Resembling a weasel, as in *Hylocichla mustelina*, the Wood Thrush, whose color may be deemed weasel-like

Muta *MOO-ta*
Mute, quiet, as in *Lagopus muta*, the Rock Ptarmigan, which has only a croaking song

Myadestes *my-a-DEST-eez*
Greek, *muia*, fly, gnat, and *edestes*, eater, as in *Myadestes townsendi*, Townsend's Solitaire

Mycteria *mik-TER-ee-a*
Greek, *mukter*, nose, snout, as in *Mycteria ibis*, the Yellow-billed Stork

Myiagra *my-AG-ra*
Greek, *muia*, fly, gnat, and *agra*, catching, seizure, as in *Myiagra atra*, the Biak Black Flycatcher

Myiarchus *my-ee-ARK-us*
Greek, *muia*, fly, gnat, and *archos*, ruler, as in *Myiarchus crinitus*, the Great Crested Flycatcher

Myioborus *my-ee-o-BOR-us*
Greek, *muia*, fly, gnat, and *borus*, eating, as in *Myioborus ornatus*, the Golden-fronted Whitestart

Myiodynastes *my-ee-o-dye-NAST-eez*
Greek, *muia*, fly, gnat, and *dynastes*, ruler, chief, as in *Myiodynastes luteiventris*, the Sulphur-bellied Flycatcher

Myioparus *my-ee-o-PAR-us*
Greek, *muia*, fly, gnat, and Latin, *parus*, a titmouse, as in *Myioparus plumbeus*, the Gray Tit-Flycatcher

Myiopsitta *my-ee-op-SIT-ta*
Greek, *muia*, fly, gnat, and Latin, *psittacus*, parrot, as in *Myiopsitta monachus*, the Monk Parakeet; not serious insect eaters, the genus name seems not to fit

Myiornis *my-ee-OR-nis*
Greek, *muia*, fly, gnat, and *ornis*, bird, as in *Myiornis ecaudatus*, the Short-tailed Pygmy-Tyrant, the smallest passerine (songbird) in the world

Myiozetetes *my-ee-o-ze-TET-eez*
Greek, *muia*, fly, gnat, and *zetetes*, a seeker, hunter, as in *Myiozetetes granadensis*, the Gray-capped Flycatcher

Myrmeciza *mer-meh-size-a*
Greek, *myrmec*, ant, and *izo*, ambush, as in *Myrmeciza goeldii*, Goeldi's Antbird

Myrmecocichla *mer-meh-ko-SICK-la*
Greek, *myrmec*, ant, and *cichla*, thrush-like bird, as in *Myrmecocichla nigra*, the Sooty Chat

Myrmornis *mir-MOR-mis*
Greek, *myrmec*, ant, and *ornis*, bird, as in *Myrmornis torquata*, the Wing-banded Antbird

Myrmotherula *mir-mo-ther-OO-la*
Greek, *myrmec*, ant, and *theras*, hunter, as in *Myrmotherula axillaris*, the White-flanked Antwren

Mystacalis *miss-ta-KAL-is*
Mustache, as in *Diglossa mystacalis*, the Moustached Flowerpiercer

Mystacea *miss-TACE-ee-a*
Greek, *mystac*, upper lip, mustache, as in *Sylvia mystacea*, Menetries's Warbler, after Édouard Ménétries, French zoologist

Myzomela *my-zo-MEL-a*
Greek, *muzo*, to suck, and *meli*, honey, as in *Myzomela erythrocephala*, the Red-headed Myzomela

Myzornis *my-ZOR-nis*
Greek, *muzo*, to suck, and *ornis*, bird, as in *Myzornis pyrrhoura*, the Fire-tailed Myzornis; its diet includes nectar and tree sap

Bird Songs and Calls

In temperate zones, breeding season for most birds is in the spring; only in the tropics is it all year. At your bird feeder, in shrubs, or in the sky you can hear birds making sounds throughout the year, but oral communication between birds is much more obvious and frequent during courtship and nesting. Songs are complex sounds typically used during breeding season to attract mates and defend territories. Calls are simple sounds usually meant to convey information such as the location of a bird, to keep a flock together, or to sound an alarm; for example, the sounds flocks of migrating geese make or the chattering around the bird feeder. Songs are produced only by birds classified as songbirds (order Passeriformes)—just over 50 percent of all the birds in the world—but not all birds of the order produce songs, such as jays and crows, for example. "Songbirds" is a common term, but those birds in the order Passeriformes are related due to their anatomic and physiologic similarities, such as the structure of the palate, the feet, and the wings, not on their singing abilities. Birds not in the Passeriformes order have calls or other vocalizations, even melodic ones, but usually just honks, quacks, grunts, wheezes, or growls.

Humans have a larynx, a set of muscles and cartilage on top of the trachea that leads to the lungs, over which air flows to produce sounds. Birds have a similar structure, a syrinx, but it is located on the bottom of the trachea, closer to the lungs and airsacs for efficient sound production. There are lots of sound-related modifications in the bird world. Swans have a long trachea that curls into the sternum, allowing the production of low frequency sounds. Oilbirds and swiftlets produce sounds similar to sonar for navigation, effective in the dark places they often inhabit. Some birds, like bushshrikes and babblers, engage in antiphonal singing: one bird of a pair sings, then the other, sounding like one bird singing.

It takes a bit of practice to identify birds by their songs or calls and it's best to learn from someone who knows them. There are recordings available on CDs and the internet to help you learn these songs, but be aware that birds, like humans, have regional accents. Different populations of songbirds have songs that differ, sometimes considerably. Sparrows from the east of North America may sound different than those in the west. So it's really best to learn songs from birds in your area.

Serinus canaria,
Atlantic Canary

The Atlantic Canary, native to the Canary Islands and Madeira, has been domesticated and bred into a variety of colors.

Some naturalists might disagree, but birds do not sing purely for enjoyment. Singing serves reproductive activities necessary for the survival of the species, for passing genes to the next generation. But it is also dangerous; singing attracts attention, and thus competitors and predators. Typically, only males sing and are attractively colored. Females rarely sing and are usually dull colored. Singing occurs usually in the breeding season. If singing were actually an expression of joy, both males and females would do it, and all year around. It may be nice to think the robin warbling his melodious song is expressing his happiness, but survival, not emotion, is the driver.

Bird song is partly genetic and partly learned. Experiments have shown that young birds, isolated from their parents, sing a song, but it's incomplete and only partially true to their species. And if they hear other songs, they incorporate parts of them. Young birds have to learn the full song by listening to their parents in the spring or summer following the year they are hatched.

Years ago, singing canaries were popular. A radio show in the US in the 1940s featured dozens of canaries singing along with classical records. There was a canary song training record you could use at home to teach your canary to sing. Their popularity led unscrupulous pet shop owners to inject both male and female canaries with testosterone, the male hormone that induces singing (and other courtship behavior). Needless to say, the new owners were disappointed when, after a few weeks, the canaries stopped singing as the hormone wore off.

For many years songs have been used as a way to identify species. Variations in the calls of similar-looking birds also provide hints to

Telophorus quadricolor,
Four-colored Bushshrike

The Four-colored Bushshrike's simple call augments its spectacular colors to attract attention.

ornithologists that these might be different species. In the US, the western and eastern populations of the Marsh Wren, *Cistothorus palustris*, for example, differ in song and are therefore being studied for a possible separation into two species. As with all other bird characteristics, new information brings new taxonomic considerations.

Cistothorus palustris,
Marsh Wren

Marsh Wrens build a domed nest with a side entrance and support it with emergent marsh plants.

N

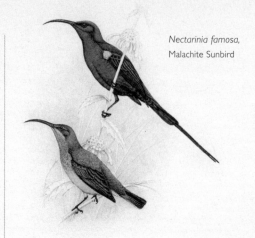

Nectarinia famosa,
Malachite Sunbird

Naevius, -a, -oides
NEE-vee-us/a/nee-vee-OID-eez
Naevus, spot or spotted, as in *Ixoreus naevius*, the Varied Thrush

Naevosa nee-VO-sa
Naevus, spot or spotted, as in *Stictonetta naevosa*, the Freckled Duck

Nahani na-HAN-eye
After P. F. Nahan, a Belgian traveler, as in *Ptilopachus nahani*, Nahan's Partridge

Nana, -nus NA-na/nus
Nanus, dwarf, as in *Acanthiza nana*, the Yellow Thornbill, with a bill not unlike a sharp thorn

Napensis na-PEN-sis
Greek, *nape*, the woods, and *-ensis*, belonging to, as in *Megascops napensis*, the Napo Screech Owl

Napothera na-po-THER-a
Greek, *nape*, the woods, and *therao*, hunt, as in *Napothera atrigularis*, the Black-throated Wren-Babbler

Natalensis na-ta-LEN-sis
Natal, South Africa, specifically the Natal, and *-ensis*, belonging to, as in *Cisticola natalensis*, the Croaking Cisticola, a bird found in Africa south of the Sahara

Natalis na-TAL-is
Birthday, as in *Ninox natalis*, the Christmas Boobook, endemic to Christmas Island in the Indian Ocean

Nativitatis na-tiv-ih-TAT-us
Nativitas, birth, as in *Puffinus nativitatis*, the Christmas Shearwater, common name from Christmas Island in the Pacific (Kiribati)

Nattererii NAT-er-er-ee-eye
After Johann Natterer, an Australian naturalist and collector, as in *Cotinga nattererii*, the Blue Cotinga

Naumanni NOY-man-eye
After Johann Naumann, German farmer and naturalist, as in *Falco naumanni*, the Lesser Kestrel

Nebouxii ne-BOUKS-ee-eye
After Adolphe Neboux, French physician and naturalist, as in *Sula nebouxii*, the Blue-footed Booby

Nebularia neb-oo-LAR-ee-a
Nebula, mist, cloudy, and *aria*, belonging to, as in *Tringa nebularia*, the Common Greenshank, with grayish-brown winter plumage

Nebulosa neb-oo-LOS-a
Nebula, mist, cloudy, as in *Strix nebulosa*, the Great Gray Owl, an allusion to its grayish color

Necropsar ne-KROP-sar
Greek, *necro*, dead, and *psar*, starling, as in *Necropsar rodericanus*, the extinct Rodrigues Starling

Nectarinia nek-tar-IN-ee-a
Greek, *nectar*, and *inus*, belonging to, as in *Nectarinia famosa*, the Malachite Sunbird

Strix nebulosa,
Great Gray Owl

Neergaardi *NER-gard-eye*
After P. Neergaard, recruiter for Witwatersrand mines, as in *Cinnyris neergaardi*, Neergaard's Sunbird

Neglecta, -us *ne-GLEK-ta/tus*
Neglected, as in *Sturnella neglecta*, the Western Meadowlark, which was for years thought to be the western population of the Eastern Meadowlark

Nehrkorni *NAIR-korn-eye*
After Adolphe Nehrkorn, German ornithologist and oologist, as in *Dicaeum nehrkorni*, the Crimson-crowned Flowerpecker

Nelsoni *NEL-son-eye*
After Edward Nelson, founding president of the American Ornithologists' Union, as in *Vireo nelsoni*, the Dwarf Vireo

Nemoricola *nem-or-ih-KO-la*
Nemus, a grove, and *colo*, dwell, as in *Gallinago nemoricola*, the Wood Snipe; Snipe from the Old Norse, *snipa*

Nemosia *ne-MO-see-a*
Nemus, a grove, as in *Nemosia pileata*, the Hooded Tanager

Neochelidon *nee-o-KEL-ih-don*
Greek, *neo*, new, and *chelidon*, swallow, as in *Neochelidon tibialis*, the White-thighed Swallow

Neochen *NEE-o-ken*
Greek, *neo*, new, and *chen*, goose, as in *Neochen jubata*, the Orinoco Goose

Neocichla *nee-o-SICK-la*
Greek, *neo*, new, and *cichla*, thrush, as in *Neocichla gutturalis*, the Babbling Starling

Neodrepanis *nee-o-dre-PAN-is*
Greek, *neo*, new, and *drepane*, a sickle, as in *Neodrepanis coruscans*, the Common Sunbird-Asity, with a sickle-shaped bill

Neomorphus *nee-o-MOR-fus*
Greek, *neo*, new, and *morphe*, form, as in *Neomorphus rufipennis*, the Rufous-winged Ground Cuckoo

Neophema *nee-o-FEEM-a*
Greek, *neo*, new, and *Euphema*, a previous genus of birds no longer used, as in *Neophema elegans*, the Elegant Parrot

Neopsittacus *nee-op-SIT-ta-kus*
Greek, *neo*, new, and Latin, *psittacus*, parrot, as in *Neopsittacus pullicauda*, the Orange-billed Lorikeet; probably reflects a newly discovered genus of parrots at the time of naming

Neospiza *nee-o-SPY-za*
Greek, *neo*, new, and *spiza*, finch, as in *Neospiza* (now *Crithagra*) *concolor*, the São Tomé Grosbeak

Neotis *nee-O-tis*
Greek, *neos*, new, and *otis*, bustard, as in *Neotis denhami*, Denham's Bustard

Nereis *NER-ee-is*
A sea god, as in *Sternula nereis*, the Fairy Tern

Nesasio *ne-SAS-ee-o*
Greek, *nesos*, island, and *asio*, little horned owl, as in *Nesasio solomonensis*, the Fearful Owl

Nesocichla *ne-so-SICK-la*
Greek, *nesos*, island, and Latin, *cichla*, thrush, as in *Nesocichla eremita*, the Tristan Thrush

Nesoctites *ne-sock-TITE-eez*
Greek, *nesos*, island, and *ktites*, inhabitants, as in *Nesoctites micromegas*, the Antillean Piculet

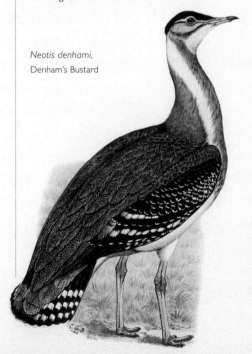

Neotis denhami, Denham's Bustard

Nesofregetta *ne-so-fre-GET-ta*
Greek, *nesos*, island, and *fregetta*, Latinized form of English frigate, as in *Nesofregetta fuliginosa*, the Polynesian Storm Petrel

Nesomimus *ne-SOM-ih-nus*
Greek, *nesos*, island, and *mimus*, mimic, as in *Nesomimus* (now *Mimus*) *trifasciatus*, the Floreana Mockingbird

Nesospiza *ne-so-SPY-za*
Greek, *nesos*, island, and *spiza*, finch, as in *Nesospiza questi*, the Nightingale Island Finch

Nesotriccus *ne-so-TRIK-kus*
Greek, *nesos*, island, and *trikkos*, a small bird, as in *Nesotriccus ridgwayi*, the Cocos Flycatcher

Nestor *NES-tor*
Hero from greek mythology, as in *Nestor meridionalis*, the New Zealand Kaka (see box), a Maori name meaning parrot

Netta *NET-ta*
Greek, *netta* or *nessa*, duck, as in *Netta erythrophthalma*, the Southern Pochard

Nestor meridionalis, New Zealand Kaka

LATIN IN ACTION

Both Kaka and *Nestor* are unusual names, but for different reasons. The New Zealand Kaka, *Nestor meridionalis*, was named by the Maori for its call and the genus name came from Greek mythology. Nestor was an Argonaut who assisted in the hunt for the centaurs and fought in the Trojan War at the age of 110. There appears to be no reason to assign the Kaka this name, but over the years, beginning with Linnaeus, names from classical mythology were occasionally chosen for birds. The specific epithet, *meridionalis*, simply means southern. The Kaka is a primitive parrot, having evolved from ancestors about five million years ago. Its brush-like tongue tip is one differentiating characteristic, allowing it to dine on nectar, as well as a variety of fruit.

Nettapus *NET-ta-pus*
Greek, *netta* or *nessa*, duck, and *pous*, foot, as in *Nettapus pulchellus*, the Green Pygmy Goose

Neumanni *NOY-man-nye*
After Oskar Neumann, a German ornithologist, as in *Urosphena neumanni*, Neumann's Warbler

Neumayer *NOY-mare*
After Franz Neumayer, Austrian botanist, as in *Sitta neumayer*, the Western Rock Nuthatch

Newelli *noo-WEL-lee-eye*
After Matthias Newell, Hawaiian missionary, as in *Puffinus newelli*, Newell's Shearwater

Newtonia, -iana *noo-TONE-ee-a/noo-tone-ee-AN-a*
After Alfred Newton, British zoologist, as in *Newtonia amphichroa*, the Dark Newtonia

Niger, -ra *NY-jer/gra*
Black, as in *Chlidonias niger*, the Black Tern

Nigrescens *nee-GRESS-sens*
Blackish, from *niger*, black, as in *Setophaga nigrescens*, the Black-throated Gray Warbler

Nigricans NEE-gri-kans
Nigrico, becoming black, from niger, black, as in Sayornis nigricans, the Black Phoebe

Nigricapillus, -ocapillus
nee-gri-ka-PIL-lus/nee-gro-ca-PIL-lus
Niger, black, and capillus, hair on the head, as in Formicarius nigricapillus, the Black-headed Antthrush

Nigricauda nee-gri-KAW-da
Niger, black, and cauda, tail, as in Myrmeciza nigricauda, Esmeraldas Antbird

Nigriceps NEE-gri-seps
Niger, black, and ceps, headed, as in Serinus nigriceps, the Ethiopian Siskin

Nigricollis nee-gri-KOL-lis
Niger, black, and collis, neck, collar, as in Grus nigricollis, the Black-necked Crane

Nigrifrons NEE-gri-fronz
Niger, black, and frons, front, forehead, as in Chlorophoneus nigrifrons, the Black-fronted Bushshrike

Nigripectus nee-gri-PEK-tus
Niger, black, and pectus, breast, as in Machaerirhynchus nigripectus, the Black-breasted Boatbill

Nigripennis nee-gri-PEN-nis
Niger, black, and penna, feather, as in Oriolus nigripennis, the Black-winged Oriole

Nigripes nee-GRIP-eez
Niger, black, and pes, foot, as in Phoebastria nigripes, the Black-footed Albatross

Nigrirostris nee-gri-ROSS-tris
Niger, black, and rostris, bill or beak, as in Andigena nigrirostris, the Black-billed Mountain Toucan

Nigrita nee-GRIT-a
Niger, black, as in Nigrita bicolor, the Chestnut-breasted Nigrita

Nigrogularis nee-gro-goo-LAR-is
Niger, black, and gularis, throat, as in Colinus nigrogularis, the Yucatan Bobwhite

Machaerirhynchus nigripectus, Black-breasted Boatbill

Nigropectus nee-gro-PEK-tus
Niger, black, and pectus, breast, as in Biatas nigropectus, the White-bearded Antshrike

Nigrorufa, -fus nee-gro-ROO-fa/fus
Niger, black, rufus, red, as in Ficedula nigrorufa, the Black-and-orange Flycatcher

Nigroventris nee-gro-VEN-tris
Niger, black, ventris, belly, as in Euplectes nigroventris, the Zanzibar Red Bishop

Nilotica, -us nee-LOT-ih-ka/us
Niloticus refers to the Nile River where Gelochelidon nilotica, the Gull-billed Tern, was first described

Ninox NY-noks
Derivation unknown, as in Ninox jacquinoti, Solomons Boobook or the Hawk-Owl

Nipalensis ni-pa-LEN-sis
After Nepal, as in Nisaetus nipalensis, the Mountain Hawk-Eagle

Nitens NI-tenz
Nitere, to shine, as in Phainopepla nitens, the Phainopepla, a silky-plumaged bird

Nitidus ni-TY-dus
Elegant, trim, gleaming, as in Buteo nitidus, the Gray-lined Hawk

Plectrophenax nivalis,
Snow Bunting

Nivalis *ni-VAL-is*
Nivis, snow, as in *Plectrophenax nivalis*, the Snow Bunting

Niveigularis *ni-vee-eye-goo-LAR-is*
Nivis, snow, and *gularis*, throat, as in *Tyrannus niveigularis*, the Snowy-throated Kingbird

Nivea *NI-vee-a*
Nivis, snow, as in *Pagadroma nivea*, the Snow Petrel

Nobilis *no-BIL-us*
Known or famous, as in *Moho nobilis*, the extinct Hawaii Oo

Nonnula *non-NOO-la*
Greek, *nonna*, nun, and *-ulus*, diminutive, as in *Nonnula ruficapilla*, the Rufous-capped Nunlet, closely related to nunbirds

Notabilis *no-TA-bil-is*
Notable, remarkable, as in *Nestor notabilis*, the Kea, common name from its call

Notata, -us *no-TA-ta/tus*
Notat, marked, as in *Meliphaga notata*, the Yellow-spotted Honeyeater

Nothocercus *no-tho-SIR-cus*
Greek, *nothos*, counterfeit, and Latin, *cerco*, tail, as in *Nothocercus julius*, the Tawny-breasted Tinamou, whose tail is virtually absent

Nothoprocta *no-tho-PROK-ta*
Greek, *nothos*, counterfeit, and *proktos*, anus or hindpart, as in *Nothoprocta ornata*, the Ornate Tinamou; counterfeit refers to the hidden tail

Nothura *no-THUR-a*
Greek, *nothos*, counterfeit, and *oura*, tail as in *Nothura maculosa*, the Spotted Nothura; counterfeit refers to the hidden tail

Notiochelidon *no-tee-o-KEL-ih-don*
Greek, *notios*, southern, and *chelidon*, swallow, as in *Notiochelidon cyanoleuca*, the Blue-and-white Swallow

Notornis *no-TOR-nis*
Greek, *notos*, the south, and *ornis*, bird, as in *Notornis* (now *Porphyrio*) *mantelli*, the Mohoau

Novaeguineae *no-vee-GWIN-ee-ee*
New Guinea, as in *Dacelo novaeguineae*, the Laughing Kookaburra

Novaehollandiae *no-vee-hol-LAND-ee-ee*
After New Holland, historical name for Australia, as in *Anhinga novaehollandiae*, the Australasian Darter

Notornis (now *Porphyrio*) *mantelli,*
Mohoau

Novaeseelandiae *no-vee-se-LAND-ee-eye*
After Zeelandia, Zeeland, Netherlands; historical name for New Zealand, as in *Aythya novaeseelandiae*, the New Zealand Scaup

Noveboracensis *no-va-bor-a-SEN-sis*
Latinized form of New York, as in *Seiurus noveboracensis*, the Northern Waterthrush

Nuchalis *noo-KAL-is*
Nucha, nape, *-alis*, belonging to, as in *Glareola nuchalis*, the Rock Pratincole, with a white collar across its nape

Nucifraga *noo-si-FRAG-a*
Nux, nut, and *frangere*, to break, as in *Nucifraga columbiana*, Clark's Nutcracker, named after the explorer William Clark

Nudiceps *NOO-di-seps*
Nudus, bare, and *ceps*, head, as in *Gymnocichla nudiceps*, the Bare-crowned Antbird

Nudicollis *noo-di-KOL-lis*
Nudus, bare, and *collis*, throat, as in *Procnias nudicollis*, the Bare-throated Bellbird

Nuditarsus *noo-di-TAR-sus*
Nudus, bare, and *tarsus*, ankle, as in *Aerodramus nuditarsus*, the Bare-legged Swiftlet

Numenius *noo-MEN-ee-us*
Greek, *noumenios*, curlew, as in *Numenius phaeopus*, the Whimbrel

Numida *noo-MID-a*
Greek, *nomas*, nomad, as in *Numida meleagris*, the Helmeted Guineafowl. These birds can wander several miles a day in search of food

Nuttallii *nut-TAL-lee-eye*
After Thomas Nuttall, English botanist and zoologist, as in *Picoides nuttallii*, Nuttall's Woodpecker

Nuttingi *NUT-ting-eye*
After Charles Nutting, American naturalist and collector, as in *Myiarchus nuttingi*, Nutting's Flycatcher

Nyctanassa *nik-ta-NAS-sa*
Greek, *nyx*, night, and *anassa*, queen, as in *Nyctanassa violacea*, the Yellow-crowned Night Heron

Nyctibius *nik-TIB-ee-us*
Greek, *nyx*, night, and, *bius*, living, as in *Nyctibius grandis*, the nocturnal Great Potoo

Glareola nuchalis,
Rock Pratincole

Nycticorax *nik-ti-KOR-aks*
Greek, *nyx*, night, and Latin, *corax*, raven, as in *Nycticorax nycticorax*, the Black-crowned Night Heron

Nycticryphes *nik-ti-KRI-feez*
Greek, *nyx*, night, and *cryptos*, hidden, as in *Nycticryphes semicollaris*, the South American Painted-snipe, a crepuscular (dusk) to somewhat nocturnal bird

Nyctidromus *nik-ti-DROM-us*
Greek, *nyx*, night, and *dromos*, runner, as in *Nyctidromus albicollis*, the Pauraque, a nocturnal bird, the common name from a Spanish transliteration of the call

Nyctiprogne *nik-tih-PROG-nee*
Greek, *nyx*, night, and Procne, a figure in Greek mythology who was turned into a swallow, as in *Nyctiprogne leucopyga*, the Band-tailed Nighthawk

Nyctyornis *nik-tee-OR-nis*
Greek, *nyx*, night, and *ornis*, a bird, as in *Nyctyornis athertoni*, the Blue-bearded Bee-eater, mistakenly described as nocturnal. Its long throat feathers give it its common name

Nymphicus *nim-FIK-us*
Nympha, nymph, and *-icus*, belonging to, derived from, as in *Nymphicus hollandicus*, the Cockatiel, named by the first Europeans to see the bird in Australia for its beauty

Nystalus *nis-TAL-us*
Greek, *nustaleos*, sleepy, as in *Nystalus maculatus*, the Caatinga Puffbird, from its lethargic behavior

Konrad Lorenz

(1903–1989)

Konrad Lorenz, zoologist, ethologist, and ornithologist, was born in Austria in 1903. His most signficant accomplishment was sharing the 1973 Nobel Prize in Physiology or Medicine with Nikolaas Tinbergen and Karl von Frisch. Lorenz studied instinctive behavior in birds and was one of the founders of the science of ethology (animal behavior). He was especially known for his explanation of imprinting, behavior that develops without any prior experience and remains for the animal's life. Lorenz credits his parents with imprinting him with a life-long love for animals and passion for birds.

Lorenz went to medical school at the University of Vienna, earned his MD, and served as an assistant professor of anatomy until 1935. Later he earned a PhD in Zoology at the same institution. Lorenz was a friend and student of renowned biologist Sir Julian Huxley, and after graduating, he met Nikolaas Tinbergen, who became a good friend and colleague. Both were interested in aspects of instinct and collaborated on studies of geese, both wild and domestic.

In 1938 Lorenz became a member of the Nazi Party and dedicated his work to the "ideas of the National Socialists." He later denied being a party member, minimized the extent of the Holocaust, and claimed he was not anti-Semitic, though his letters show otherwise.

In 1940 Lorenz became a professor of psychology at the University of Königsberg. A year later he was conscripted into the German Army as a clinical psychologist. Near the end of the war in 1944 he was sent to the Russian front and was held as a prisoner of war for four years. He continued his studies even as a prisoner and kept a pet starling as a companion.

In 1958, Lorenz was employed at the Max Planck Institute for Behavioral Physiology and remained there until 1973, the year he received his shared Nobel Prize. Besides the Nobel Prize, Lorenz received other honors, including the Austrian Decoration for Science and Art (1964) and the Gold Medal of the Humboldt Society (1972).

Lorenz was the author of several books. Perhaps his best known are *King Solomon's Ring* and *On Aggression*, both written for a general audience. In the first book Lorenz asserts that his power to communicate with animals is comparable to King Solomon's. *On Aggression* argues that all animals, especially males, are aggressive as a way to gain and protect resources.

Here, Lorenz, right, walks with Nikolaas Tinbergen, fellow Nobel Prize laureate who also worked with birds.

"Truth in science can be defined as the working hypothesis best suited to open the way to the next better one."

Konrad Lorenz

Coloeus monedula,
Western Jackdaw

Jack, from late fourteenth-century England, refers to one of lower class, and daw, comes from an old English name of the bird.

A couple of his influences were German ornithologist Oskar Heinroth and American biologist Charles Whitman, who studied birds without any preconception of what they were supposed to do or be. Heinroth and Whitman influenced Lorenz's ideas on instinct as it applied to avian social behavior. He captured a Western Jackdaw, *Coloeus monedula*, tamed it, noted its behavior, and eventually established a colony of the birds in the family home. His first published scientific paper (1927) described the social behavior of jackdaws.

Considered the father of ethology, Lorenz wrote numerous books on the subject. Working with Nikolaas Tinbergen, he developed the idea of an innate releasing mechanism, a process by which a stimulus evokes a response when the connection between the two is inborn, to explain instinctive behaviors. To prove this hypothesis, they studied the behavior of birds and co-authored a paper on the rolling behavior of Greylag Geese, *Anser anser*. At the sight of a single egg outside of the nest, the Greylag Goose will roll the egg back to the others with its beak. If the egg is removed, the animal continues to engage in egg-rolling behavior, pulling its head back as if there were still an egg there.

Although a controversial figure, Lorenz received many significant awards in ornithology and several honorary doctorates.

Anser anser,
Greylag Goose

The Greylag Goose is the ancestor of all domestic geese, including the white ones for the dinner table.

Sasia ochracea, White-browed Piculet

Oatesi *OATS-eye*
After Eugene Oates, British civil servant in India, as in *Hydrornis oatesi*, the Rusty-naped Pitta

Oberholseri *ob-ber-HOLT-ser-eye*
After Harry Oberholser, American ornithologist, as in *Empidonax oberholseri*, the American Dusky Flycatcher

Obscurus, -a *ob-SKUR-us/a*
Obscure, as in *Dendragapus obscurus*, the Dusky Grouse; obscure refers to the dull color

Obsoletus, -a *ob-so-LEE-tus/ta*
Plain, ordinary, as in *Salpinctes obsoletus*, the Rock Wren

Obtusa *ob-TOO-sa*
Obtusus, dull, blunt, as in *Vidua obtusa*, the Broad-tailed Paradise Whydah

Occidentalis *ok-si-den-TAL-is*
Occidere, to fall, as the sun in the west, hence western, as in *Larus occidentalis*, the Western Gull

Occipitalis *ok-si-pi-TAL-is*
Occiput, back of the head, as in *Dendrocitta occipitalis*, the Sumatran Treepie, with a white nape

Occulta *ok-KUL-ta*
Occultus, hidden, concealed, as in *Pterodroma occulta*, the Vanuatu Petrel; the species name reflects that little is known about this bird

Oceanica, -us *o-see-AN-ih-ka/kus*
Oceanus, ocean, as in *Ducula oceanica*, the Micronesian Imperial Pigeon, which lives on islands of the Pacific

Oceanicus, -a *o-see-AN-ih-kus/ka*
Greek, *oceanic*, as in *Oceanites oceanicus*, Wilson's Storm Petrel

Oceanites *o-see-an-EYE-teez*
Greek, god of the sea, as in *Oceanites gracilis*, Elliot's Storm Petrel

Oceanodroma *o-see-an-o-DROM-a*
Oceanus, ocean, and *dromos*, running, as in *Oceanodroma furcata*, the Forked-tail Storm Petrel, which "runs" on the ocean's surface

Ocellata, -um, -us *o-sel-LAT-a/um/us*
Ocellus, eye, and *-ata*, having, as in *Meleagris ocellata*, the Ocellated Turkey, with eyespots on the tail

Ochotensis *o-ko-TEN-sis*
Greek, refers to the Sea of Okhotsk, as in *Locustella ochotensis*, Middendorff's Grasshopper Warbler, which has an Asian distribution

Ochracea *o-KRACE-ee-a*
Ochra, pale yellow, as in *Sasia ochracea*, the White-browed Piculet

Ochraceiceps *o-krace-ee-EYE-seps*
Ochra, pale yellow, and *ceps*, headed, as in *Hylophilus ochraceiceps*, the Tawny-crowned Greenlet

Ochraceifrons *o-krace-ee-EYE-fronz*
Ochra, pale yellow, and *frons*, forehead, as in *Grallaricula ochraceifrons*, the Ochre-fronted Antpitta

Ochraceiventris *ok-ra-see-eye-VEN-tris*
Ochra, pale yellow, and *ventris*, belly, as in *Leptotila ochraceiventris*, the Ochre-bellied Dove

Ochraceus, -a *ok-RACE-ee-us/a*
Pale yellow, as in *Contopus ochraceus*, the Ochraceous Pewee

Ochrocephala *ok-ra-se-FAL-a*
Ochra, pale yellow, and *cephala*, head, as in *Amazona ochrocephala*, the Yellow-crowned Amazon

Ochrogaster ok-kro-GAS-ter
Ochra, pale yellow, and Greek, *gaster*, stomach, as in *Penelope ochrogaster*, the Chestnut-bellied Guan

Ochrolaemus o-kro-LEE-mus
Ochra, pale yellow, and Greek, *laemus*, throat, as in *Automolus ochrolaemus*, the Buff-throated Foliage-gleaner

Ochthoeca ak-THO-ee-ka
Greek, *okhthos*, mound, and *oikos*, home, as in *Ochthoeca* (now *Silvicultrix*) *frontalis*, the Crowned Chat-Tyrant, which forages from mounds of moss and dead bamboo

Ochthornis ak-THOR-nis
Greek, *okhthos*, mound, and *ornis*, bird, as in *Ochthornis littoralis*, the Drab Water Tyrant, which often sits upon exposed roots or piles of debris along riverbanks

Ocreatus, -ta o-kree-AH-tus/ta
Ocrea, leg covering, as in *Ocreatus underwoodii*, the Booted Racket-tail

Ocularis a-koo-LAR-is
Oculus, eye, of the eye, as in *Ploceus ocularis*, the Spectacled Weaver

Oculocincta o-koo-lo-SINK-ta
Oculus, eye, and *cinctus*, girdle, crown as in *Oculocincta squamifrons*, the Pygmy White-eye

Ocyalus o-see-AL-us
Greek, Ocale, one of the mythical Amazons, as in *Ocyalus latirostris*, the Band-tailed Oropendola

Ocyceros o-see-SER-os
Greek, *oxy*, sharp, and *keras*, horn, as in *Ocyceros birostris*, the Indian Gray Hornbill

Ocyphaps O-see-faps
Greek, *oxy*, sharp, and *phaps*, dove, as in *Ocyphaps lophotes*, the Crested Pigeon

Odontophorus o-don-toe-FOR-us
Greek, *odontos*, tooth, and *phoreus*, bearer, as in *Odontophorus melanotis*, the Black-eared Wood Quail, with serrated maxilla

Odontorchilus o-don-tor-KIL-us
Greek, *odontos*, tooth, and *cheilos*, lip as on a pitcher, as in *Odontorchilus cinereus*, the Tooth-billed Wren

Odontospiza o-don-to-SPY-za
Greek, *odontos*, tooth, and *spiza*, finch, as in *Odontospiza griseicapilla*, the Gray-headed Silverbill, which has a very toothlike bill

Oena o-EE-na
Greek, *oinas*, meaning pigeon, referring to the color of nearly ripened grapes, as in *Oena capensis*, the Namaqua Dove, from the Nama people of southern Africa

Oenanthe o-ee-NAN-thee
Greek, *oine*, vine, and *anthus*, blossom, as in *Oenanthe monticola*, the Mountain Wheatear, so called because of its spring appearance (after migration) at the time when wine vines blossom

Oglei O-gul-eye
After M. J. Ogle, British surveyor and collector, as in *Stachyris oglei*, the Snowy-throated Babbler

Oidemia oy-DEE-mee-a
Greek, swelling, as in *Oidemia* (now *Melanitta*) *nigra*, the Common Scoter, with a swollen bill

Ocreatus underwoodii, Booted Racket-tail

Oleagineus o-lee-a-JIN-ee-us
Of the olive, as in *Mionectes oleagineus*, the Ochre-bellied Flycatcher

Olivacea, -um, -us o-liv-ACE-see-a/um/us
Olive-green colored, as in *Spinus olivacea*, the Olivaceous Siskin

Olivii o-LIV-ee-eye
After Edmund Olive, Australian naturalist and collector, as in *Turnix olivii*, the Buff-breasted Buttonquail

Olor O-lor
Swan, as in *Cygnus olor*, the Mute Swan

Olrogi OL-rog-eye
After Claes Olrog, Swedish ornithologist, as in *Cinclodes olrogi*, Olrog's Cinclodes

Omissa o-MIS-sa
Missing, omitted, as in *Foudia omissa*, the Forest Fody, which is sometimes considered as part of another species and thus overlooked

Oncostoma on-ko-STOM-a
Greek, *onco*, mass, body size, and *stoma*, mouth, as in *Oncostoma olivaceum*, the Southern Bentbill

Onychognathus on-ee-kog-NA-thus
Greek, *onycho-*, claw, nail, and *gnathos*, jaw, as in *Onychognathus neumanni*, Neumann's Starling, with a heavy curved bill

Onychoprion on-ee-ko-PRY-on
Greek, *onux*, claw, nail, and *prion*, saw, as in *Onychoprion lunatus*, the Spectacled Tern; the bird's middle claw has small serrations

Onychorhynchus on-ee-ko-RINK-us
Greek, *onux*, claw, nail, and Latin, *rhynchus*, bill, as in *Onychorhynchus coronatus*, the Amazonian Royal Flycatcher, with a hooked bill

Ophrysia o-FRIS-ee-a
Greek, *ophrys*, eyebrow, as in *Ophrysia superciliosa*, the likely extinct Himalayan Quail, with a white supercilium (eyebrow)

Opisthocomus o-pis-tho-KO-mus
Greek, *opistho*, behind, backward, and *comus*, hair, as in *Opisthocomus hoazin*, the Hoatzin, whose head is topped by a spiky crest

Opisthoprora o-pis-tho-PRO-ra
Greek, *opistho*, behind, backward, and *prora*, prow, as in *Opisthoprora euryptera*, the Mountain Avocetbill, with an unusual bill for a hummingbird

Oreomanes fraseri, Giant Conebill

Oporornis o-por-OR-nis
Greek, *opora*, autumn, and *ornis*, bird, as in *Oporornis agilis*, the Connecticut Warbler

Orchesticus or-KES-ti-kus
Orchestra, a place for dancers to perform, and *icas*, belonging to (i. e. dancer), as in *Orchesticus abeiliei*, the Brown Tanager; though rarely observed, its courtship display may explain the name

Oreocharis or-ee-o-KAR-is
Oros, mountain, and Greek, *charis*, graciousness, gratitude, beauty, as in *Oreocharis arfaki*, the Tit Berrypecker, an attractive bird that typically live in mountain forest above 2,200 meters

Oreoica or-ee-O-ik-a
Oros, mountain, and *-ica*, belonging to, as in *Oreoica gutturalis*, the Crested Bellbird, which lives in mountain habitats

Oreomanes or-ee-o-MAN-eez
Oros, mountain, and *manes*, spirit, as in *Oreomanes fraseri*, the Giant Conebill; the bird's spirit is in the mountains (the Andes)

Oreomystis or-ee-o-MIS-tis
Oros, mountain, and *mysticus*, mystic, as in *Oreomystis bairdi*, the Akikiki, which lives only in the highest rainforest habitats of Kauai, Hawaii

Oreonympha or-ee-o-NIM-fa
Oros, mountain, and *nympha*, nymph, goddess of the mountains, as in *Oreonympha nobilis*, the Bearded Mountaineer

Oreophasis *or-ee-o-FAY-sis*
Greek, *oros*, mountain, and Latin, *phasianus*, pheasant, as in *Oreophasis derbianus*, the Horned Guan, which lives in mountain habitats

Oreophilus *or-ee-o-FIL-us*
Greek, *oros*, mountain and *philos*, love, loving, as in *Buteo oreophilus*, the Mountain Buzzard

Oreophylax *or-ee-o-FYE-laks*
Greek, *oros*, mountain, and *phylax*, guard, protect, as in *Oreophylax* (now *Asthenes*) *moreirae*, the Itatiaia Spinetail; Itatiaia is a Brazilian municipality

Oreopsittacus *or-ee-op-SIT-ti-kus*
Greek, *oros*, mountain, and Latin, *psittacus*, parrot, as in *Oreopsittacus arfaki*, the Plum-faced Lorikeet, which lives in mountain habitats

Oreornis *or-ee-OR-nis*
Greek, *oros*, mountain, and *ornis*, bird, as in *Oreornis chrysogenys*, the Orange-cheeked Honeyeater, which lives in mountain habitats

Oreortyx *or-ee-OR-tiks*
Greek, *oros*, mountain, and Latin *ortyx*, quail, as in *Oreortyx pictus*, the Mountain Quail (see box)

Oreoscoptes *or-ee-o-SCOP-teez*
Greek, *oros*, mountain, and *scoptes*, mimic, as in *Oreoscoptes montanus*, the Sage Thrasher, or the Mountain Mockingbird

Oreoscopus *or-ee-o-SKO-pus*
Greek, *oros*, mountain, and *scopos*, watcher, as in *Oreoscopus gutturalis*, the Fernwren

Oreothraupis *or-ee-o-THRAW-pis*
Greek, *oros*, mountain, and *thraupis*, a small bird, as in *Oreothraupis arremonops*, the Tanager Finch

> ### LATIN IN ACTION
>
> The only member of its genus, the Mountain Quail (*Oreortyx pictus*) lives mainly in the mountains of the western US. *Pictus*, from the Latin, is an adjective meaning painted or embroidered, and one look at this bird and you will know why. The gray of the chest, nape, and top of the head contrasts with the rusty-red and white belly and flanks. The long feathers of its crest remind one of an exclamation point. Typically found at elevations of 2,3000 to 990 feet (700 to 300 meters), they will migrate altitudinally to avoid snow-covered ground in the winter and move as far as 20 miles (32 kilometers) between seasons to remain in suitable habitat.

Orientalis *or-ee-en-TAL-is*
Of the east, as in *Merops orientalis*, the Green Bee-eater

Oriolus, -lia *or-ee-O-lus/lee-a*
Aureolus, gold, golden, as in *Oriolus flavocinctus*, the Green Oriole

Ornata, -tus *or-NA-ta/tus*
Ornate, as in *Urocissa ornata*, the Sri Lanka Blue Magpie

Ornithion *or-NITH-ee-on*
Greek, *ornis*, bird, and Latin, *-ion*, being, as in *Ornithion inerme*, the White-lored Tyrannulet

Oroaetus *or-o-EE-tus*
Greek, *oros*, mountain, and *aetos*, eagle, as in *Oroaetus* (now *Spizaetus*) *isidori*, the Black-and-chestnut Eagle

Oreortyx pictus, Mountain Quail

Ortalis *or-TAL-is*
Greek, chicken, as in *Ortalis vetula*, the Plain Chachalaca, an onomatopoetic name for the call

Orthogonys *or-tho-GON-is*
Greek, *orthos*, straight, and *genys*, jaw, as in *Orthogonys chlorichterus*, the Olive-green Tanager

Orthonyx *or-THON-iks*
Greek, *orthos*, straight, and *onux*, claw, as in *Orthonyx novaeguineae*, the Papuan Logrunner

Orthopsittaca *or-thop-SIT-tak-a*
Greek, *orthos*, straight, and Latin, *psittaca*, parrot, as in *Orthopsittaca manilatus*, the Red-bellied Macaw

Orthorhyncus *or-tho-RINK-us*
Greek, *orthos*, straight, and *rhynchos*, bill, as in *Orthorhyncus cristatus*, the Antillean Crested Hummingbird, with a straight bill, indicative of generalist hummingbirds

Orthotomus *or-tho-TOE-mus*
Greek, *orthos*, straight, and *tomus*, piece, section, as in *Orthotomus sutorius*, the Common Tailorbird, which pierces the edges of a leaf, rolls it over and sews the edges together to form the base for a nest

Ortygospiza *or-ti-go-SPY-za*
Greek, *ortux*, quail, and *spiza*, finch, as in *Ortygospiza atricollis*, the Quailfinch

Ortyxelos *or-tiks-EL-os*
Greek, *ortux*, quail, and *elos*, low ground, as in *Ortyxelos meiffrenii*, the Quail-plover

Oryzivorus, -a *or-riz-ih-VOR-us/a*
Oryza, rice, and *vorus*, eat, swallow, as in *Dolichonyx oryzivorus*, the Bobolink; refers to the bird's liking for (cultivated) rice and other grains

Ossifragus *os-si-FRAY-gus*
Ossis, bone, and *frangere*, break, as in *Corvus ossifragus*, the Fish Crow. Its diet includes fish, the bones of which it breaks whilst eating

Ostralegus *os-tra-LEG-us*
Greek, *ostreon*, oyster, and *lego*, gather, as in *Haematopus ostralegus*, the Eurasian Oystercatcher

Otidiphaps *o-TI-di-faps*
Greek, *otis*, bustard, and *phaps*, wild pigeon, as in *Otidiphaps nobilis*, the Pheasant Pigeon

Otis *O-tis*
Greek, *otis*, bustard, as in *Otis tarda*, the Great Bustard

Otus *O-tus*
Little horned owl, as in *Otus asio*, the Eastern Screech Owl

Oustaleti *oo-sta-LET-eye*
After Emile Oustalet, French zoologist, as in *Cinnyris oustaleti*, Oustalet's Sunbird

Oxylabes *aks-ih-LAY-beez*
Greek, *oxus*, sharp, and Latin, *labe*, slip, slide, as in *Oxylabes madagascariensis*, the White-throated Oxylabes, with a sharp bill

Oxypogon *aks-ee-PO-gon*
Greek, *oxus*, sharp, and, *pogon*, beard, as in *Oxypogon guerinii*, the Bearded Helmetcrest

Oxyruncus *aks-ee-RUN-kus*
Greek, *oxus*, sharp, and Latin, *rhynchus*, bill, as in *Oxyruncus cristatus*, the Sharpbill

Oxyura, -us *aks-ee-OO-ra/rus*
Greek, *oxus*, sharp, and *oura*, tail, as in *Oxyura jamaicensis*, the Ruddy Duck, a member of the stiff-tailed duck family

Oxylabes madagascariensis,
White-throated Oxylabes

OTUS

This genus is known as the scops owls due to a confusing taxonomic dispute in the eighteenth century when *Scops* was used as the genus name. The most recently discovered species, the Serendib Scops Owl (*O. thilohoffmanni*), was found in Sri Lanka in 2006.

Latin for eared owl,= the genus name *Otus* probably was a reference to the feather tufts that resemble ears. *Otus* has 63 species, including Scops-Owls and Screech Owls. Most are small, have cryptic plumage with spots or streaks, and conspicuous ear tufts. These tufts have no hearing function but their position may indicate the mood of a bird or serve to break up the outline of the owl as it sits motionless in a tree. Owls have incredibly good hearing and can precisely locate prey because their external fleshy ears are different in shape from each other and positioned asymmetrically, thus enabling owls to triangulate the source of the sound because the sound waves are altered as they enter the ear.

Owls' large ears are matched by their large eyes and keen eyesight. A large owl's eyes are as big as a human's. The eyes are tubular and fixed in their skull by a ring of bony plates, but the birds can turn their heads 270 degrees and have excellent stereoscopic vision. They have a large number of light sensitive cells in their retina, and can theoretically see by the light of a match held 0.6 miles (1 kilometer) away.

Otus gurneyi,
Giant Scops Owl

Contrary to popular belief, owls can see perfectly well in the daytime.

After locating their prey with their two outstanding senses, they fly almost noiselessly over the landscape to capture it, thanks to their special feather structure. The edges of their flight feathers are frayed, with a soft, downy covering to the feathers, so when the owls fly, the feathers emit only low frequency sounds that their prey, like a mouse, cannot detect.

Owls have zygodactyl feet, two toes facing forward and two back, but they can reverse one toe so that they have three forward and one back. This may be an adaptation to make perching and catching prey easier. Owls eat invertibrates, small mammals, birds, and reptiles, which are swallowed whole or in big chunks. The food makes its way down to the muscular part of the stomach, the gizzard, where it is ground. The digestible portion continues down the digestive tract but the undigestible parts—bones, feathers, fur— are compacted into an "owl pellet" and regurgitated.

Otus owls are found everywhere on earth except Antarctica and Australia and are most common in Asia.

Otus silvicola and *Otus alfredi*,
Wallaces's Scops Owl and
Flores Scops Owl

Common Names

A scientific name clearly designates a particular bird species, tells you something about the bird's relationship to other bird species, and usually provides a decent description of the bird. Common names are less useful for the former, because they vary so much from country to country, but apt for the latter. The obvious advantage common names have is their much easier pronunciation and spelling. And for English names, at least, the International Ornithologists' Union has recommended English common names and set standards for spelling and construction. There are still many problems with common names due to their long history and local variations.

There are common names like the Zitting Cisticola, *Cisticola juncidis*, Plain Chachalaca, *Ortalis vetula*, Kea, *Nestor notabilis*, or the Phainopepla, *Phainopepla nitens*, that tell you nothing about the bird. There are birds named after people, like Abert's Towhee, *Melozone aberti*, or Salvin's Chuckwill, *Antrostomus salvini*, and those names aren't much more useful, although there has been a trend toward eliminating the personal name from the common name; hence *Oenanthe phillipsi*, Phillip's Wheatear, became the Somali Wheatear, Salvin's Chuckwill is now the Tawny-colored Nightjar, and Meyer's Sicklebill, *Epimachus meyeri*, got renamed the Brown Sicklebill. Helpfully, there are many common names that closely reflect the scientific name such as the Green-headed Oriole, *Oriolus chlorocephalus*, and *Chen caerulescens*, the Snow or Blue Goose. But the opposite is also true; *Ploceus melanogaster*, meaning black-bellied weaver, has the common name of Black-billed Weaver; and *Myrmeciza melanoceps*, the White-shouldered Antbird, whose scientific name means Black-headed Antbird. Sometimes the common name includes part of the scientific name such as *Oxylabes madagascariensis*, the White-throated Oxylabes, and *Rhabdornis mysticalis*, the Stripe-headed Rhabdornis.

Some common names have been changed for clarification or simplification: the Celebes Bearded Bee-eater, *Meropogon forsteni*, has been shortened to the Celebes Bee-eater; the Lance-billed Monklet has become the Lanceolated Monklet, *Micromonacha lanceolata*; and Mayr's Streaked Honeyeater is now simply Mayr's Honeyeater, *Ptiloprora mayri*.

Some common names come from the bird's calls, such as the Plain Chachalaca, Kea, and New Zealand Kaka, *Nestor meridionalis*; these are onomatopoeic words as they phonetically imitate the call. Common names also come from the local language, such as Western Capercaillie, *Tetrao urogallus*, from the Scottish Gaelic; the Akohekohe, *Palmeria dolei*; and Tui, *Prosthemadera novaeseelandiae*, from the Maori.

Odd things happen as well. Some common names mislead the reader; for example, you might think a Western Meadowlark, *Sturnella neglecta*, is a lark, when it is actually a blackbird. The Red-bellied Woodpecker, *Melanerpes carolinus*, has a not very obvious pinkish wash on its belly and the neck ring of the

Meropogon forsteni,
Purple-bearded Bee-eater

A typical bee-eater, the Celebes or Purple-bearded Bee-eater is very colorful and captures large insect prey in flight.

Ortalis vetula,
Plain Chachalaca

Like many birds with odd common names, the Plain Chachalaca gets its name from its call.

and Feral Pigeon. The Redbird became the Northern Cardinal, *Cardinalis cardinalis*; the Sea-swallow was renamed the Common Tern, *Sterna hirundo*; and in the US the Linnet became the House Finch, *Haemorhous mexicanus*. The Goldcrest, *Regulus regulus*, was once known as the Woodcock Pilot as it arrived in the spring at the same time as the Eurasian Woodcock, *Scolopax rusticola*. It was believed that Goldcrests were too small to migrate on their own and so arrived tucked into the feathers of the Woodcock.

Although both common and scientific naming are dynamic processes, there are standards for both. And it is certainly nice to be able to speak of the Coppersmith Barbet instead of the tongue-twisting *Megalaima haemacephala*!

Ring-necked Duck, *Aythya collaris*, is almost impossible to see in the field. Dove and pigeon are used pretty interchangeably, the difference being that the former is of Anglo-Saxon origin and the latter French. Same for the Germanic-derived name heron and the French-derived egret appellation.

There are lots of regional and country differences. Loons, as they are known in America, are referred to as divers in Europe. And while Americans call *Buteo* species hawks, the British call them buzzards. There are also spelling differences like grey vs. gray and colour vs. color; the International Ornithologists' Union standards lean toward using the British spelling.

Common names, like the scientific names, keep evolving. The European Robin, *Erithacus rubecula*, was originally called Redbreast, then Robin Redbreast, Ruddock, Robinet, and sometimes the English Robin. The familiar Rock Dove, *Columbia livia* has been called the Rock Pigeon, Carrier Pigeon, Common Pigeon, Homing Pigeon,

Buteo solitarius,
Hawaiian Hawk

Like most hawks, the endangered Hawaiian Hawk lives and hunts alone except during the breeding season.

P

Pachycare *pak-ih-KAR-ee*
Greek, *pakhus*, thick, and *care*, head, as in *Pachycare flavogriseum*, the Goldenface, a member of the Pachycephalidae family, once known as "thickheads" because of their dumpy bodies and large heads

Pachycephala, -cephalopsis *pak-ih-se-FAL-a/pak-ih-se-fal-OP-sis*
Greek, *pakhus*, thick, and Latin, *cephala*, head, as in *Pachycephala olivacea*, the Olive Whistler, with a large head

Pachycoccyx *pak-ih-KOK-siks*
Greek, *pakhus*, thick, and *coccyx*, cuckoo, as in *Pachycoccyx audeberti*, the Thick-billed Cuckoo

Pachyptila *pak-ip-TIL-a*
Greek, *pakhus*, thick, and *ptilon*, feather, as in *Pachyptila desolata*, the Antarctic Prion, common name from Greek *prioni*, saw, referring to the serrated beak

Pachyramphus *pak-ih-RAM-fus*
Greek, *pakhus*, thick, and Latin, *rhamphus*, beak, as in *Pachyramphus viridis*, the Green-backed Becard, from French *becarde*, beak

Pachyrhyncha *pak-ih-RINK-a*
Greek, *pakhus*, thick, and Latin, *rhynchus*, beak, as in *Rhynchopsitta pachyrhyncha*, the Thick-billed Parrot

Pacifica, -us *pa-SIF-ik-a/us*
Of the Pacific Ocean, as in *Gavia pacifica*, the Pacific Loon or Diver

Padda *PAD-da*
Refers to paddy fields of rice, as in *Padda* (now *Lonchura*) *oryzivora*, the Java Sparrow

Palmeria dolei, Akohekohe

Pagodroma *pa-go-DROME-a*
Greek, *pagos*, cold, and *dromos*, running, as in *Pagodroma nivea*, the Snow Petrel

Pagophila *pa-go-FIL-a*
Greek, *pagos*, cold, and *philos*, loving, as in *Pagophila eburnea*, the Ivory Gull, which lives in the high Arctic

Pallasii *pal-LASS-ee-eye*
After Peter Pallas, German zoologist, as in *Cinclus pallasii*, Pallas's or the Brown Dipper

Palliseri *PAL-li-ser-eye*
After Edward and F. H. Palliser, collectors in Sri Lanka, as in *Elaphrornis palliseri*, Palliser's or the Sri Lanka Bush Warbler

Palmeria *pal-MAIR-ee-a*
After Henry Palmer, collector on Hawaii, as in *Palmeria dolei*, the Akohekohe

Pandion *PAN-ee-on*
After a King of Athens whose daughters were turned into birds, as in *Pandion haliaetus*, the Western Osprey

Panurus *pan-OO-rus*
Panu, all, and *oura*, tail, as in *Panurus biarmicus*, the Bearded Reedling, with its small body and long tail

Cinclus pallasii, Pallas's or Brown Dipper

Parabuteo *par-a-BOO-tee-o*
Greek, *para*, like or near, and *buteo*, hawk, as in *Parabuteo unicinctus*, Harris's Hawk. Audubon named this bird after his friend Edward Harris

Paradigalla *par-a-di-GAL-la*
Greek, *para*, like or near, and Latin, *gallus*, chicken, as in *Paradigalla brevicauda*, the Short-tailed Paradigalla

Paradisaea *par-a-DEES-ee-a*
Greek, *paradeisos*, a park or garden, as in *Paradisaea minor*, the Lesser Bird-of-paradise

Paradoxornis *par-a-doks-OR-nis*
Greek, *paradoxos*, strange, amazing and *ornis*, bird, as in *Paradoxornis guttaticollis*, the Spot-breasted Parrotbill

Pardalotus *par-da-LO-tus*
Greek, *pardalotos*, spotted, as in *Pardalotus punctatus*, the Spotted Pardalote

Parkeri *PAR-ker-eye*
After Theodore Parker, American ornithologist, as in *Cercomacra parkeri*, Parker's Antbird

Parotia *par-OT-ee-a*
Greek, *parotis*, the parotid gland, as in *Parotia sefilata*, the Western Parotia, *parotia* probably referring to head plumes behind the ear

Parula *pa-ROO-la*
Diminutive of *parus*, titmouse or little tit, as in *Parula* (now *Setophaga*) *americana*, the Northern Parula

Paradoxornis guttaticollis, Spot-breasted Parrotbill

LATIN IN ACTION

From Middle English *poucock*, the peacock, native to southeast Asia, is known worldwide. *Pavo cristatus* describes the crested peacock, officially known as the Indian Peafowl. The fan-shaped crest on the head is impressive and distinctive, but the iridescent blue body and the spectacular train of tail feathers that spreads into a giant fan of iridescence with eyespots seem to beg for description. An interesting myth still given credence today in India is that the female peahen fertilizes her eggs by drinking the tears of the male peacock.

Parus *PA-rus*
Parus, titmouse, as in *Parus major*, the Great Tit; tit perhaps from Norse *tita*, small bird

Passer *PAS-ser*
Sparrow, as in *Passer domesticus*, the House Sparrow

Passerculus *pas-ser-COO-lus*
Diminutive of *passer*, sparrow, as in *Passerculus sandwichensis*, the Savannah Sparrow

Passerella *pas-ser-EL-la*
Diminutive of *passer*, sparrow, as in *Passerella iliaca*, the Fox Sparrow

Passerherbulus *pas-ser-her-BOO-lus*
Passer, sparrow, and *herbulus*, little herb, as in *Passerherbulus* (now *Ammodramus*) *caudacutus*, Saltmarsh Sparrow

Passerina *pas-ser-ee-na*
Diminutive of *passer*, sparrow, as in *Passerina caerulea*, the Blue Grosbeak

Pavo *PA-vo*
Peacock, as in *Pavo cristatus*, the Indian or Blue Peacock (see box)

Pealii *PEEL-ee-eye*
After Titian Peale, an American naturalist and artist, as in *Erythrura pealii*, the Fiji Parrotfinch

PASSER

True sparrows belong to the genus *Passer* (*PAS-ser*), Latin for sparrow. There are about 27 species in the genus, most of which are distributed in the warmer climates of southern Eurasia and Africa. They are mainly ground-dwelling seedeaters but will take insects when raising young. Since the genus name means sparrow, it is up to the specific name to describe the bird, but that does not always work. The scientific name *P. flaveolus*, meaning golden, doesn't aptly describe the Plain-backed Sparrow any better than the common name. *P. italiae*, the Italian Sparrow, at least provides a locale, and *P. rutilans*, meaning reddish, auburn, works well for the Russet Sparrow.

Because the *Passer* sparrows are seedeaters, they have special adaptations in their jaws, hard palate, and tongue that help them manipulate and open seeds. The sparrows hold large seeds transversely and crack them open against the hard palate.

Sparrows enjoy bathing by standing in water puddles and ducking their heads under water. They also dust bathe by scratching out a depression in the soil and spreading their wings out. Bathing is followed by intensive preening that not only helps to rid the feathers of parasites, but supports the social unit. After bathing, the birds often gather to roost or sing.

Fully 17 of 27 species of *Passer* sparrows nest in or on human-made structures. Certainly the most widespread, abundant, and well-known of all sparrows is *P. domesticus*, the House Sparrow, completely adapted to human habitation. Once known as the English Sparrow, the House Sparrow has been introduced and has spread around the world. Its success over the last century has led to its being considered a pest. Introduced into New York in 1852, they now are found from northern Canada to Panama. During World War I in England, Sparrow Clubs were created to rid the countryside of House Sparrows, and in the 1960s Mao Tse-tung declared a war on House and Eurasian Tree Sparrows (*P. montanus*) and had millions killed, which is reputed to have led to severe famine a few years later as insects decimated the crops.

House Sparrows may still be deemed a pest by some, but because they are abundant and easy to keep in captivity, they have been used in over 5,000 scientific studies. An increase in pesticide use may be the cause of the huge decline in this species in Europe.

Passer domesticus, House Sparrow

Passer ammodendri, Saxaul Sparrow

The Saxaul Sparrow, native to central Asia, is one of the larger *Passer* sparrows. Its head markings make it very distinctive.

Pectoralis *pek-to-RA-lis*
Pectoro-, breast, chest, as in *Euphonia pectoralis*, the Chestnut-bellied Euphonia

Pedionomus *ped-ee-o-NO-mus*
Greek, *pedion*, plain, field, and *nomos*, a home, as in *Pedionomus torquatus*, the Plains-wanderer

Pelagodroma *pel-a-go-DRO-ma*
Greek, *pelagos*, sea, and *dromos*, runner, as in *Pelagodroma marina*, the White-faced Storm Petrel, for its habit of pattering its feet on the sea surface

Pelecanoides *pel-eh-kan-OY-deez*
Greek, *pelekan*, pelican, and *oides*, resembles, as in *Pelecanoides urinatrix*, the Common Diving Petrel, which resembles a pelican

Pelecanus *pel-eh-KAN-us*
Greek, *pelekan*, pelican, as in *Pelecanus conspicillatus*, the Australian Pelican

Pelzelnii *pel-ZEL-nee-eye*
After August von Pelzeln, Austrian ornithologist, as in *Tachybaptus pelzelnii*, the Madagascar Grebe

Penelope *pen-EL-o-pee*
Greek, Penelopeia, feminine name, as in *Penelope albipennis*, the White-winged Guan

Penicillatus *pen-ih-sil-LA-tus*
Penicullus, brush, as in *Phalacrocorax penicillatus*, Brandt's Cormorant; the breeding male has tufts of feathers on its head

Pennula *pen-NOO-la*
Diminutive of *penna*, wing, as in *Pennula* (now *Porzana*) *sandwichensis*, the extinct Hawaiian Rail, a small-winged flightless bird

Perdicula *per-di-KOO-la*
Perdix, a partridge, and *cula*, little, as in *Perdicula erythrorhyncha*, the Painted Bush Quail

Perdix *PER-diks*
Perdix, a partridge, as in *Perdix perdix*, the Gray Partridge

Pericrocotus *per-ih-kro-KO-tus*
Greek, *peri*, around, and *crocotus*, golden-yellow, as in *Pericrocotus roseus*, the Rosy Minivet

Perisoreus *pe-ri-SOR-ee-us*
Unclear derivation but may refer to bird's habit of storing food, as in *Perisoreus infaustus*, the Siberian Jay

Perissocephalus *pe-ris-so-se-FAL-us*
Greek, *perissos*, strange, excessive, and Latin, *cephala*, head, as in *Perissocephalus tricolor*, the Capuchinbird, which resembles a monk with its bald face and collar of feather

Pernis *PER-nis*
Greek, corruption of *pternis*, bird of prey, as in *Pernis ptilorhynchus*, the Crested Honey Buzzard

Peronii *per-OWN-ee-eye*
After Francois Péron, French explorer and naturalist, as in *Geokichla peronii*, the Orange-sided Thrush

Personata, -us *per-son-AH-ta/tus*
Persona, masked, as in *Coracina personata*, the Wallacean Cuckooshrike

Petiti *PE-ti-tye*
After Louis Petit, French naturalist, as in *Campephaga petiti*, Petit's Cuckooshrike

Petrochelidon *pe-tro-KEL-ih-don*
Petra, rock, and Greek, *chelidon*, swallow, as in *Petrochelidon nigricans*, the Tree Martin, which nests in tree cavities and rock crevices

Pelecanus conspicillatus, Australian Pelican

Petroica *pe-TRO-ee-ka*
Petra, rock, and *-icus*, belonging to, as in *Petroica australis*, the New Zealand Robin, which uses rocks as hunting perches

Petronia *pe-TRO-nee-a*
Greek, *petronius*, of a rock, as in *Petronia* (now *Gymnoris*) *superciliaris*, the Yellow-throated Petronia, which was mistakenly thought to be a rock dweller

Peucedramus *poy-se-DRA-mus*
Greek, *peuke*, pine, and *dromos*, runner, as in *Peucedramus taeniatus*, the Olive Warbler, which feeds around pine trees

Phacellodomus *fa-sel-lo-DO-mus*
Greek, *phakelos*, package, and *domos*, house, as in *Phacellodomus ruber*, the Greater Thornbird, which constructs a complex nest of twigs

Phaenicophaeus *fee-ni-KO-fee-us*
Greek, *phoiniko*, crimson, and *phaeinos*, shining, as in *Phaenicophaeus* (now *Zanclostomus*) *javanicus*, the Red-billed Malkoha

Phaeochroa *fee-o-KRO-a*
Greek, *phaeo*, dusky, and *chroa*, color, as in *Phaeochroa cuvierii*, the Scaly-breasted Hummingbird

Phaeornis *fee-OR-nis*
Greek, *phaeo*, brown, dark, dusky, and *ornis*, bird, as in *Phaeornis* (now *Myadestes*) *obscurus*, the Omao

Phaethon *FAY-eh-thon*
In Greek mythology Phaethon was the son of Helios, the sun, as in *Phaethon lepturus*, the White-tailed Tropicbird

Phainoptila melanoxantha, Black-and-yellow Phainoptila

LATIN IN ACTION

The scientific name *Phainoptila melanoxantha*, meaning shining-feather black-yellow (bird), aptly describes the Black-and-yellow Phainoptila, a resident of the cloud forests of Costa Rica and part of Panama. It is one of three genera of silky flycatchers in the Ptiliogonatidae family. The term flycatcher refers to aerial insect-eating behavior, but in fact it is not related to the Muscicapidae (flycatcher) or Tyrannidae (tyrant) families of flycatchers. The bird's diet consists almost exclusively of fruits; insect-catching is a rare event.

Phaethornis *fay-eh-THOR-nis*
See *Phaethon*, above, and Greek, *ornis*, bird, as in *Phaethornis ruber*, the Reddish Hermit

Phainopepla *fay-no-PEP-LA*
Greek, *phaeinos*, shining, and *peplos*, robe or cloak, as in *Phainopepla nitens*, the Phainopepla, named for its silky plumage

Phainoptila *fay-nop-TIL-a*
Greek, *phaeinos*, shining, and *ptilon*, feather, as in *Phainoptila melanoxantha*, the Black-and-yellow Phainoptila (see box)

Phalacrocorax *fal-a-kro-KOR-aks*
Greek, *phalakros*, bald, and *corus*, raven, as in *Phalacrocorax brasilianus*, the Neotropic Cormorant

Phalaenoptilus *fal-ee-nop-TIL-us*
Greek, *phalaina*, moth, and *ptilon*, feather, as in *Phalaenoptilus nuttallii*, the Common Poorwill, with soft grayish-brownish plumage

Phalaropus *fal-a-RO-pus*
Greek, *phalaris*, coot, and *pous*, foot, as in *Phalaropus tricolor*, Wilson's Phalarope, with partial webbing on its feet, as seen in coots

Phalcoboenus *fal-ko-BAY-nus*
Greek, *phalkon*, falcon, and *baino*, walking, as in *Phalcoboenus australis*, the Striated Caracara, often seen walking on the ground

Phaps *FAPS*
Greek, *phaps*, dove or pigeon, as in *Phaps chalcoptera*, the Common Bronzewing

Pharomachrus *fa-ro-MAK-rus*
Greek, *pharos*, cloak, and *macros*, long, large, as in *Pharomachrus auriceps*, the Golden-headed Quetzal

Phasianus *fay-see-AN-us*
From *phasiana*, a reference to the River Phasis (now Rioni, Georgia), where the Common Pheasant, *Phasianus colchicus*, was once common

Pheucticus *FOIK-ti-kus*
Greek, *pheuktikos*, shy or timid, as in *Pheucticus chrysogaster*, the Southern Yellow Grosbeak

Philacte *fil-AK-tee*
Greek, *philos*, like or love, and *akte*, shore, as in *Philacte* (now *Chen*) *canagica*, the Emperor Goose

Philepitta *fil-eh-PIT-ta*
Greek, *philos*, love, and *pitta*, from the Telugu (an Indian language) word meaning small bird, as in *Philepitta castanea*, the Velvet Asity

Philetairus *fil-eh-TARE-us*
Greek, *philos*, love, and *hetairos*, companion, as in *Philetairus socius*, the Sociable Weaver

Phillipsi *FIL-lips-eye*
After E. Lort Phillips, British big game hunter, as in *Oenanthe phillipsi*, the Somali Wheatear

Philomachus *fil-o-MAK-us*
Greek, *philos*, like or love, and *makhe*, fight, battle, as in *Philomachus pugnax*, the Ruff, named for its aggressive behavior during lekking

Phleocryptes *flee-o-KRIP-teez*
Greek, *phleos*, an aquatic plant, and *cryptus*, hidden, as in *Phleocryptes melanops*, the Wren-like Rushbird

Phloeoceastes *flo-ee-o-see-steez*
Greek, *phloios*, tree bark, and *keazo*, split, cleave, as in *Phloeoceastes* (now *Campephilus*) *robustus*, the Robust Woodpecker, a bark splitter

Phodilus *fo-DIL-us*
Greek, *phos*, light, and *deilos*, afraid, fear, as in *Phodilus prigoginei*, the Congo Bay Owl

Phoeniconaias *foy-ni-KO-nye-as*
Greek, *phoinikos*, red, and *naias*, water nymph, as in *Phoeniconaias minor*, the Lesser Flamingo

Philepitta castanea,
Velvet Asity

Phoenicoparrus *foy-ni-ko-PAR-rus*
Greek, *phoinikos*, red, and Latin *parra*, ominous bird, as in *Phoenicoparrus andinus*, the Andean Flamingo

Phoenicopterus *foy-ni-KOP-ter-us*
Greek, *phoinikos*, red, and *pteron*, wing, as in *Phoenicopterus roseus*, the Greater Flamingo, whose common name derives from the Latin *flamma*, flame

Phoenicurus *foy-ni-KOO-rus*
Greek, *phoinikos*, red, and *oura*, tail, as in *Phoenicurus frontalis*, the Blue-fronted Redstart

Phrygilus *fri-JIL-us*
Greek, *phrugilos*, a bird, and *-icus*, belonging to, as in *Phrygilus atriceps*, the Black-hooded Sierra Finch. From Aristophanes, who called anyone from Phyriga (part of modern Turkey) a *phrygilus*, or finch

Phyllastrephus *fil-la-STREF-us*
Greek, *phyllon*, leaf, and *strepho*, twist, as in *Phyllastrephus terrestris*, the Terrestrial Brownbul, which lives in dense, dry thickets of savannah or acacia

Phylloscopus *fil-lo-SKOPE-us*
Greek, *phyllon*, leaf, and *skopeo*, seek, as in *Phylloscopus borealis*, the Arctic Warbler, which spends much time feeding in the leafy canopies of trees

Phytotoma *fy-to-TO-ma*
Greek, *phuton*, plant, and *tomos*, a cut-off piece, as in *Phytotoma rutila*, the White-tipped Plantcutter

PHOENICOPTERUS

The Red Queen in *Alice in Wonderland* used flamingos as croquet mallets, presumably because of their long necks—longer than any other bird relative to body size—and their upside-down mallet-shaped heads with large bills. There are three genera of flamingos: Three species of *Phoenicopterus* (red-feathered), one species of *Phoeniconaias* (red naiad), and two species of *Phoenicoparrus* (red water bird). The word flamingo derives from the Spanish *flamenco*, meaning flame-colored.

The flamingo's bill allows it to feed like baleen whales; in both animals the inner part of the jaw is covered with numerous lamellae and the tongue, moving over these lamellae like a piston, filters out microorganisms from the water. These microorganisms, animal and plant plankton, contain carotinoids, protein pigments that give the birds their color. Because the concentration of these pigments in their food sources vary, the color of flamingo populations and individuals varies from whitish to red.

The six species of flamingo are found primarily in the southern hemisphere, but also in Spain, the Caribbean, and coasts from Arabia east to India. They are very social birds, often found in flocks numbering thousands of birds. At times Kenya's Lake Nakuru has over a million birds. The warm alkaline lake provides the birds abundant algae, which itself is dependent on the droppings of the birds for nutrients.

Phoenicopterus roseus, Greater Flamingo

The names of the birds are pretty straightforward. *P. roseus* (Latin, rose-colored) is the Greater Flamingo; *P. ruber* (Latin, red) is the American Flamingo; *P. chilensis* is the Chilean Flamingo; *P. minor* is the Lesser Flamingo; and *P. andinus* is the Andean Flamingo. *P. jamesi*, James's Flamingo, was named after British tycoon Harry Berkley James, who sponsored an expedition that discovered the bird in Bolivia in 1886.

Flamingos build a volcano-shaped nest and usually lay one egg on the top. The gray-plumaged chick is born with a straight red bill that develops the adult curve later. Filter feeding by the adults poses a problem for feeding the young, but a special adaptation solves the problem. The crop, an expanded part of the esophagus, produces a protein-rich secretion that both the male and female feed to the young. It is called flamingo milk and is similar to pigeon milk.

Phoeniconaias minor, Lesser Flamingo

The three flamingo species are very similar in structure and habits and are only differentiated by minor differences in their feeding mechanism.

Pica, -us *PIKE-a/us*
Latin for magpie, as in *Pica pica*, the Eurasian Magpie

Picoides *pi-KOY-deez*
Picus, woodpecker, and *eidos*, shape, likeness, as in *Picoides arcticus*, the Black-backed Woodpecker

Piculus *pi-KOO-lus*
Picus, woodpecker, and *-ulus*, diminutive, little, as in *Piculus litae*, the Lita Woodpecker

Picumnus *pik-KUM-nus*
In Roman mythology, Picumnus was a god of fertility, as in *Picumnus exilis*, the Golden-spangled Piculet

Pileata, -us *pil-ee-AH-ta/tus*
Pileatus, capped, as in *Piprites pileata*, the Black-capped Piprites

Pinaroloxias *pin-a-ro-LOKS-ee-as*
Greek, *pinaros*, dirty, and *loxos*, slanting, as in *Pinaroloxias inornata*, the Cocos Finch

Pinarornis *pin-a-ROR-nis*
Greek, *pinaros*, dirty, and *ornis*, bird, as in *Pinarornis plumosus*, the Boulder Chat, presumably because of its dirty, blackish color

Pinguinus *pin-GWIN-us*
Welsh, *pen*, head, and *gwyn*, white, as in *Pinguinus impennis*, the extinct Great Auk, named for its similarity to penguins

Pinicola *pin-ih-KO-la*
Pinus, pine, and *cola*, inhabitant, as in *Pinicola enucleator*, the Pine Grosbeak

Pipilo *PIP-il-o*
Pipo, to chirp, as in *Pipilo* (now *Melozone*) *fusca*, the Canyon Towhee

Pipra *PIP-ra*
Greek, *pipra*, bird, as in *Dixiphia pipra*, the White-crowned Manakin

Pipreola *pip-ree-O-la*
Greek, *pipra*, bird, and *-ola*, diminutive, as in *Pipreola formosa*, the Handsome Fruiteater

Piprites *pip-RITE-eez*
Greek, *pipra*, bird, and *-ites*, belonging, as in *Piprites chloris*, the Wing-barred Piprites

Piranga *pi-RANG-ga*
A Brazilian municipality, as in *Piranga rubra*, the Summer Tanager

Pitangus *pi-TANG-us*
Tupi (native Brazilian), *pitangua*, meaning large flycatcher, as in *Pitangus sulphuratus*, the Great Kiskadee or Kiskadee Flycatcher

Pithecophaga *pith-eh-ko-FAY-ga*
Greek, *pithekos*, ape, and *phagein*, to eat, as in *Pithecophaga jefferyi*, the Philippine or Monkey-eating Eagle

Pitohui *pit-o-HOO-ee*
"Pitohui" is the sound made after a human tastes and immediately rejects the poisonous bird, as in *Pitohui dichrous*, the Hooded Pitohui

Pitta *PIT-ta*
East Indian word for a small bird, as in *Pitta sordida*, the Hooded Pitta

Pittasoma *pit-ta-SO-ma*
Pitta, East Indian word for a small bird, and Greek, *soma*, body, as in *Pittasoma rufopileatum*, the Rufous-crowned Antpitta

Pityriasis *pit-ih-RYE-a-sis*
Greek, *pituron*, warts on the head, as in *Pityriasis gymnocephala*, the Bornean Bristlehead (*Pityriasis* is a flaking of the skin in humans)

Platalea *plat-AL-ee-a*
Greek, *platy*, flat, as in *Platalea minor*, the Black-faced Spoonbill

Piranga rubra, Summer Tanager

LATIN IN ACTION

Podargus ocellatus, the "lazy-footed bird with eyespots," is the Marbled Frogmouth. The order Caprimulgiformes to which this bird belongs alludes to the idea that with their big mouths the birds of this order could suckle on goats' teats, hence the old name "goatsucker." Frogmouths are found across southeast Asia to Australia. Although they appear to have small beaks, their mouths are huge, so they not only devour insects, but small lizards, mice, birds, and snakes. Their feet are weak so the birds lie horizontally across a branch during the day, hidden by their cryptic plumage. They lay their eggs in a tree branch, nestless.

Podargus ocellatus, Marbled Frogmouth

Plateni PLAT-en-eye
After Carl Platen, a German doctor and collector, as in *Dasycrotapha plateni*, the Mindanao Pygmy Babbler

Platycercus plat-ih-SIR-kus
Greek, *platy*, flat, and *cercus*, tail, as in *Platycercus adscitus*, the Pale-headed Rosella

Platypsaris plat-ip-SAR-is
Greek, *platy*, flat, and *psar*, starling, as in *Platypsaris* (now *Pachyramphus*) *aglaiae*, the Rose-throated Becard

Platyrinchus plat-ih-RINK-us
Platys, flat, and *rhynchus*, bill, as in *Platyrinchus coronatus*, the Golden-crowned Spadebill

Plautus PLAW-tus
Plautus, flat-footed, as in *Plautus* (now *Alle*) *alle*, the Little Auk or Dovekie, a bird that is clumsy on land

Plectrophenax plek-tro-FEN-aks
Greek, *plectron*, spur or cock's spur, and *phenax*, imposter, as in *Plectrophenax nivalis*, the Snow Bunting, with a long claw on the hind toe

Plectropterus plek-TROP-ter-us
Greek, *plectron*, spur or cock's spur, and *pteron*, wing, as in *Plectropterus gambensis*, the Spur-winged Goose, with a spur on each wing, used for attacking other water birds

Plectorhyncha plek-to-RINK-a
Greek, *plectron*, spur or cock's spur, and *rhynchos*, bill, as in *Plectorhyncha lanceolata*, the Striped Honeyeater, with a fine pointed bill

Plegadis ple-GA-dis
Greek, *plegas*, scythe, sickle, as in *Plegadis falcinellus*, the Glossy Ibis, with a sickle-shaped bill

Pleskei PLES-kee-eye
After Theodor Pleske, Russian zoologist and geographer, as in *Locustella pleskei*, Styan's or Pleske's Grasshopper Warbler

Plocepasser plo-see-PAS-ser
Greek, *plokeus*, weaver, and *passer*, sparrow, as in *Plocepasser superciliosus*, the Chestnut-crowned Sparrow-Weaver

Ploceus PLO-see-us
Greek, *plokeus*, weaver, as in *Ploceus luteolus*, the Little Weaver

Plumbeus, -a PLUM-bee-us/a
Leaden, the color of lead, as in *Myioparus plumbeus*, the Gray Tit-Flycatcher

Pluvialis, -anus ploo-vee-AL-is/ploo-vee-AN-us
Pluvia, rain, as in *Pluvialis squatarola*, the Gray Plover; plover from Old French *plovier*, meaning rainbird, as migratory flocks arrived at the rainy season

Podargus po-DAR-gus
Greek, *pous*, foot, and *argos*, lazy, slow, as in *Podargus ocellatus*, the Marbled Frogmouth, named for its large froglike gape (see box)

Podica *PO-di-ka*
Greek, *pous*, foot, and *-icus*, belonging to, as in *Podica senegalensis*, the African Finfoot, with lobed toes to aid propulsion through water

Podiceps *PO-di-seps*
Podex, buttocks, vent, and *pes*, foot, as in *Podiceps major*, the Great Grebe; refers to the feet being located under the bird's "vent" (rear end)

Podilymbus *po-di-LIM-bus*
Podex, buttocks, vent, and *colymbus*, swimming pool or bath, as in *Podilymbus podiceps*, the Pied-billed Grebe, pied referring to its white bill with a black spot

Poephila *po-eh-FIL-a*
Greek, *poa*, grass, and *philos*, love, as in *Poephila personata*, the Masked Finch

Pogoniulus *po-gon-ee-OO-lus*
Greek, *pogon*, beard, and *-ulus*, diminutive, as in *Pogoniulus simplex*, the Green Tinkerbird, so-called because of its "tink-tink-tink" call. Little beard refers to the heavy facial bristles characteristic of the family, although not this particular bird

Polihierax *po-lee-HY-er-aks*
Greek, *polios*, gray, and *hierax*, hawk, as in *Polihierax insignis*, the White-rumped Falcon

Poliocephala, -us *po-lee-o-se-FAL-a/us*
Greek, *polios*, gray, and Latin, *cephala*, head, as in *Chloephaga poliocephala*, the Ashy-headed Goose

Polioptila *po-lee-op-TIL-a*
Greek, *polios*, gray, and *ptilon*, feather, as in *Polioptila dumicola*, the Masked Gnatcatcher

Polyborus *pol-ee-BOR-us*
Greek, *poly*, many and *boros*, devouring, as in *Polyborus* (now *Caracara*) *cheriway*, the Northern Crested Caracara, which eats a wide variety of live and dead foods

Polyplectron *pol-ee-PLEK-tron*
Greek, *poly*, many and *plektron*, spur or cock's spur, as in *Polyplectron chalcurum*, the Bronze-tailed Peacock-Pheasant; the male bird has two spurs on its leg

Polysticta *pol-ee-STIK-ta*
Greek, *poly*, many and *stiktos*, dotted, dappled, as in *Polysticta stelleri*, Steller's Eider; although the bird has few spots, they are large and obvious

Pomatorhinus *po-ma-to-RYE-nus*
Greek, *poma*, a cover, and *rhinos*, nose, as in *Pomatorhinus gravivox*, the Black-streaked Scimitar Babbler, scimitar from the long downcurved bill

Pooecetes *poo-eh-SEE-teez*
Greek, *poe*, grass, and *oiketes*, inhabitant, as in *Pooecetes gramineus*, the Vesper Sparrow

Porphyrio *por-FEER-ee-o*
Water hen, as in *Porphyrio porphyrio*, the Purple Swamphen

Porphyrolaema *por-feer-o-LEE-ma*
Greek, *porphyros*, purple, and *laimos*, throat, as in *Porphyrolaema porphyrolaema*, the Purple-throated Cotinga, *cotinga* from Brazilian Tupi language

Porphyrospiza *por-feer-o-SPY-za*
Greek, *porphyros*, purple, and *spiza*, finch, as in *Porphyrospiza caerulescens*, the Blue Finch

Portoricensis *por-tor-ih-SEN-sis*
After Puerto Rico, as in *Spindalis portoricensis*, the Puerto Rican Spindalis. Common name apparently a misspelling and combination of other words

Porzana *por-ZAN-a*
Italian, *porzana*, the name of the bird in Italy, as in *Porzana porzana*, the Spotted Crake, common name from Old Norse *kraka*, after the call

Porphyrio porphyrio, Purple Swamphen

Premnoplex prem-NO-pleks
Greek, *premnon*, tree trunk, and *plexus*, knitting, interweaving, as in *Premnoplex tatei*, the White-throated Barbtail, which weaves its nest around a bough

Pretrei PRET-tre-eye
After Jean Pretre, French artist and illustrator, as in *Amazona pretrei*, the Red-spectacled Amazon

Prigoginei pri-go-JEEN-eye
After Alexandre Prigogine, Russian/Belgian naturalist, as in *Cinnyris prigoginei*, the Prigogine's Double-collared Sunbird

Princeps PRIN-seps
First, chief, first to take, as in *Accipiter princeps*, the New Britain Goshawk

Prinia PRIN-ee-a
Javanese, *prinya*, as in *Prinia polychroa*, the Brown Prinia

Prionochilus pry-on-o-KIL-us
Greek, *prion*, sawlike, and *kheilos*, a rim, edge, as in *Prionochilus maculatus*, the Yellow-breasted Flowerpecker, which has a beak with a serrated edge

Prionops PRY-o-nops
Greek, *prion*, sawlike, and *opsis*, appearance, as in *Prionops plumatus*, the White-crested Helmetshrike; refers to the fringed fleshy wattles around the eyes

Probosciger pro-BOS-si-ger
Proboscis, nose, snout and *ger*, bear, carry, as in *Probosciger aterrimus*, the Palm Cockatoo

Procellaria pro-sel-LAR-ee-a
Procella, storm, and *-arius*, referring to, as in *Procellaria parkinsoni*, the Black Petrel, a bird associated with storms

Procelsterna pro-sel-STER-na
Procella, storm, and *sterna*, tern, as in *Procelsterna albivitta*, the Gray Noddy; terns are often associated with storms

Procnias PROC-nee-as
Procne in Greek mythology, daughter of Pandion who was turned into a swallow, as in *Procnias nudicollis*, the Bare-throated Bellbird

Prodotiscus pro-doe-TISS-kus
Prodo, disclose, and *-iscus*, diminutive, as in *Prodotiscus regulus*, the Brown-backed Honeybird, which discloses the source of honey

Progne PROG-nee
Latin for Procne in Greek mythology, daughter of Pandion who was turned into a swallow, as in *Progne elegans*, the Southern Martin

Promerops PRO-mer-ops
Pro, for, and *merops*, bee-eater, as in *Promerops cafer*, the Cape Sugarbird

Prosthemadera pros-theme-a-DER-a
Greek, *prosthema*, an addition, and *dera*, neck or throat, as in *Prosthemadera novaeseelandiae*, the Tui, with a tuft of white feathers on its neck

Protonotaria pro-to-no-TAR-ee-a
Protos, first, and *notarius*, scribe, as in *Protonotaria citrea*, the Prothonotary Warbler. High-ranking notaries, prothonotaries, of the Byzantine Empire wore yellow robes, the color of the bird

Prunella proo-NEL-la
Corruption of *bruneus*, brown, as in *Prunella collaris*, the Alpine Accentor, with a basic brown color. *Accentor* comes from *ad*, with, and *cantor*, sing

Przewalskii she-VAL-skee-eye
After Nikolai Mikhaylovich Przhevalsky, Russian Cossack naturalist, as in *Paradoxornis przewalskii*, Przevalski's Nuthatch

Psalidoprocne sal-ih-doe-PROK-nee
Greek, *psalis*, a knife or shears, and Procne from Greek mythology, daughter of Pandion who was turned into a swallow, as in *Psalidoprocne nitens*, the Square-tailed Saw-wing

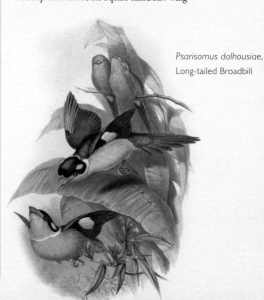

Psarisomus dalhousiae, Long-tailed Broadbill

Psaltriparus *sal-tri-PAR-us*
Psaltria, female lute player, as in *Psaltriparus minimus*, American Bushtit, with a high-pitched call

Psarisomus *sar-ih-SO-mus*
Greek, *psaros*, speckled, and *soma*, body, as in *Psarisomus dalhousiae*, the Long-tailed Broadbill

Psarocolius *sar-o-KOL-ee-us*
Greek, *psar*, starling, and *kolios*, a kind of woodpecker, as in *Psarocolius viridis*, the Green Oropendola

Pseudocalyptomena *soo-doe-kal-ip-toe-MEN-a*
Pseudo, false, Greek, *calypto*, hidden, and *mena*, moon, as in *Pseudocalyptomena graueri*, Grauer's Broadbill, considered by the namer to resemble a species belonging to the genus *Calyptomena* only slightly

Pseudochelidon *soo-doe-KEL-ih-don*
Pseudo, false, and *chelidon*, swallow, as in *Pseudochelidon eurystomina*, the African River Martin

Pseudodacnis *soo-soe-DAK-nis*
Pseudo, false, and *dacnis*, unknown Egyptian bird, such as *Pseudodacnis* (now *Dacnis*) *hartlaubi*, the Turquoise Dacnis

Pseudonestor *soo-doe-NES-tor*
Pseudo, false, and *nestor*, referring to some New Zealand parrots, as in *Pseudonestor xanthophrys*, the Maui Parrotbill

Psittacula *sit-ta-KOO-la*
Psittacus, a parrot, and *-ula*, diminutive, as in *Psittacula krameri*, the Rose-ringed or Ring-necked Parakeet

Psittacus *SIT-ta-kus*
Psittacus, a parrot, as in *Psittacus erithacus*, the Gray Parrot

Psittirostra *sit-ti-ROSS-tra*
Psittacus, parrot, and *rostrum*, beak, as in *Psittirostra psittacea*, the extinct Ou

Psophia *so-FEE-a*
Greek, *psophos*, noise, as in *Psophia viridis*, the Dark-winged Trumpeter

Psophodes *so-FO-deez*
Greek, *psophodes*, noisy, as in *Psophodes nigrogularis*, the Western Whipbird; referring to its active, lively, continual singing

Pteridophora *ter-ih-do-FOR-a*
Greek, *pteridon*, a fern, and *phoreo*, to bear, as in *Pteridophora alberti*, the King of Saxony Bird-of-paradise, with two long feathers on its head

Pseudocalyptomena graueri, Grauer's Broadbill

Pterocles *TER-o-kleez*
Greek, *pteron*, wings, as in *Pterocles coronatus*, the Crowned Sandgrouse

Pterodroma *ter-o-DROM-a*
Greek, *pteron*, wing, and *dromos*, runner, as in *Pterodroma inexpectata*, the Mottled Petrel

Pteroglossus *ter-o-GLOS-sus*
Greek, *pteron*, wing, and *glossa*, tongue, as in *Pteroglossus viridis*, the Green Aracari, which has a long, fringed tongue; Aracari from the Brazilian Tupi language

Pteropodocys *ter-o-po-DOE-sis*
Greek, *pteron*, wing, and *pous*, foot, as in *Pteropodocys* (now *Coracina*) *maxima*, the Ground Cuckooshrike, supposedly nearly as fast on the ground as it is in the air

Pteroptochos *ter-op-TOE-kos*
Greek, *pteron*, wing, and *ptokhos*, begging, as in *Pteroptochos castaneus*, the Chestnut-throated Huet-huet, named after the call

Ptilinopus *til-in-O-pus*
Greek, *ptilon*, feather, and *pous*, foot, as in *Ptilinopus cinctus*, the Banded Fruit Dove, with feather-covered tarsus

Ptiliogonys *tili-o-GON-is*
Greek, *ptilon*, feather, and *gonys*, knee, as in *Ptiliogonys cinereus*, the Gray Silky-flycatcher, with feather-covered knees

Ptilonorhynchus *til-o-no-RINK-us*
Greek, *ptilon*, feather, and Latin, *rhynchos*, beak, as in *Ptilonorhynchus violaceus*, the Satin Bowerbird, whose bill is partly covered by feathers

Ptiloprora til-o-PRO-ra
Greek, *ptilon*, feather, and *prora*, front, prow, as in *Ptiloprora erythropleura*, the Rufous-sided Honeyeater

Ptiloris til-OR-is
Greek, *ptilon*, wing, and *oris*, mouth, as in *Ptiloris magnificus*, the Magnificent Riflebird; the underside and top of the beak is partially feathered

Ptychoramphus ti-ko-RAM-fus
Greek, *ptyx*, folded, and *ramphos*, beak, as in *Ptychoramphus aleutica*, Cassin's Auklet; the bill looks as if it has been compressed and folded

Pucherani poo-cher-AN-eye
After Jacques Pucheran, French zoologist, as in *Melanerpes pucherani*, the Black-cheeked Woodpecker

Pucrasia poo-KRAS-ee-a
Nepalese, *pukras*, as in *Pucrasia macrolopha*, the Koklass Pheasant; both genus and common name derive from its call

Puffinus puf-FINE-us
Middle English, *poffin*, for the carcasses of Manx Shearwaters (used as food), as in *Puffinus gravis*, the Great Shearwater

Pulchella, -us pul-KEL-la/lus
Beautiful little, as in *Lacedo pulchella*, the Banded Kingfisher

Pulcher PUL-ker
Beautiful, as in *Lamprotornis pulcher*, the Chestnut-bellied Starling

Pulcherrima, -us pul-ker-REE-ma/mus
Pulcherrimus, very beautiful, as in *Megalaima pulcherrima*, the Golden-naped Barbet

Pulchra PUL-kra
Pulcher, beautiful, as in *Macgregoria pulchra*, MacGregor's Honeyeater

Punctatus, -a, -um punk-TAT-us/a/um
Punctum, dot, spot, as in *Falco punctatus*, the Mauritius Kestrel, with heavily spotted plumage

Pusilla, -lus poo-SIL-la/lus
Pusillis, very small, as in *Emberiza pusilla*, the Little Bunting

Pycnonotus pik-no-NO-tus
Greek, *pychnos*, strong, thick, and *notos*, back, as in *Pycnonotus nigricans*, the African Red-eyed Bulbul

Lacedo pulchella, Banded Kingfisher

Pycnoptilus pik-nop-TIL-us
Greek, *pychnos*, strong, thick, and *ptiolon*, feather, as in *Pycnoptilus floccosus*, the Pilotbird, a plump bird, implying thick plumage

Pygoscelis pi-gos-SEL-is
Greek, *puge*, the rump, and *skelos*, leg, as in *Pygoscelis papua*, the Gentoo Penguin; refers to the thick tail that brushes the ground as if it were a third leg

Pyriglena py-ri-GLEN-a
Greek, *pyr*, fire, and *glene*, eyeball, as in *Pyriglena atra*, the Fringe-backed Fire-eye

Pyrocephalus pye-ro-se-FAL-us
Greek, *pyr*, fire, and Latin, *cephala*, head, as in *Pyrocephalus rubinus*, the Vermilion Flycatcher

Pyrrhula pir-ROO-la
Greek, *pyrrhos*, fire-colored, as in *Pyrrhula pyrrhula*, the Eurasian Bullfinch, for the red plumage of the male

Pyrrhuloxia pir-roo-LOKS-ee-a
Greek, *pyrrhos*, fire-colored, and *loxos*, slanting, crosswise, as in *Pyrrhuloxia* (now *Cardinalis*) *sinuatus*, the Pyrrhuloxia. Its reddish color and compact, heavy curved bill accounts for the name

Pyrrhura pir-ROO-ra
Greek, *pyrrhos*, fire-colored, as in *Pyrrhura frontalis*, the Maroon-bellied Parakeet

Q

Quadragintus kwa-dra-JIN-tus
Forty, as in *Pardalotus quadragintus*, the Forty-spotted Pardalote

Quadribrachys kwa-dri-BRAK-is
Quadri-, four, and *brachium*, arm, as in *Alcedo quadribrachys*, the Shining-blue Kingfisher; actually means four toes

Quadricinctus kwa-dri-SINK-tus
Quadri-, four, and *cinctus*, surround, encircle, as in *Pterocles quadricinctus*, the Four-banded Sandgrouse (see box)

Quadricolor kwa-dri-KO-lor
Quadri-, four, and *color*, color in appearance, as in *Dicaeum quadricolor*, the Cebu Flowerpecker

Quelea KWEL-lee-a
After an African native name, as in *Quelea quelea*, the Red-billed Quelea

Querquedula kwer-kweh-DOO-la
Kind of duck that makes a sound like querquedula, as in *Anas querquedula*, the Garganey, which derives from the Latin *gargala*, tracheal artery

Quinticolor kwin-ti-KO-lor
Quint-, five, and *color*, visible color, as in *Capito quinticolor*, the Five-colored Barbet

---

LATIN IN ACTION

The Four-banded (*quadricinctus*) Sandgrouse inhabits the central region of Africa, from east to west, in open, savanna-like habitats. The male is larger and more colorful, but both sexes are cryptically colored to blend into the background. The bands on the chest and abdomen of the strongly marked male camouflage the bird as they break up its outline; this is called "disruptive coloration," and is often found on ground-dwelling birds. The male has specially constructed feathers (unique to species of sandgrouse) on its belly that absorb water so that it can carry water to his chicks from distant waterholes.

Pterocles quadricinctus,
Four-banded Sandgrouse

Quiscalus, -a kwis-KAL-us/a
Quis, who, and *qualis*, of what kind, as in *Quiscalus quiscula*, the Common Grackle

Quitensis kwin-TEN-sis
After Quito, Ecuador, as in *Grallaria quitensis*, the Tawny Antpitta

Quoyi KWOY-eye
After Jean Quoy, French naturalist, as in *Cracticus quoyi*, the Black Butcherbird

Capito quinticolor,
Five-colored Barbet

ALEXANDER F. SKUTCH
(1904–2004)

Born near Baltimore, Maryland, in 1904, Alexander Skutch began developing his passion for nature when his family moved to a farm in the country. He went on to study botany, receiving a doctorate degree from John Hopkins University in 1928. After graduation he sailed from New York to Panama to study banana diseases, but quickly became fascinated by the birds of the New World Tropics, as they nested all over his primitive field station.

While Skutch pursued his botanical work in Honduras, Guatemala, and Costa Rica, his interest in ornithology continued to grow. He financed his bird-watching studies by collecting tropical plants for museums in America and Europe, spending several seasons in the forests and mountains of Guatemala, Honduras, and Costa Rica. He found a perfect bird study location in a remote forested valley near San Isidro del General in Costa Rica. In 1941 he bought 178 acres of land and built a house there, naming it *Finca Los Cusingos* after the local name for the Fiery-billed Aracari, *Pteroglossus frantzii*, a relative of the toucan.

A lifelong vegetarian, Skutch grew corn, yucca, and other crops, and, without running water until the 1990s, bathed and drank from the nearest stream. He believed in "treading lightly on the mother Earth," and his longevity demonstrates that he thrived on this simple lifestyle. He married Pamela Lankester, daughter of the English naturalist Sir Charles Lankester, in 1950, and with their adopted son Edwin, stayed at *Finca Los Cusingos* for the rest of his life, although his pristine forest became an island in the middle of banana and coffee plantations.

He was a prolific naturalist/writer, publishing over 40 books, mostly about birds but also on environmental philosophy. In 1983 he produced *Birds of Tropical America*, and with Gary Stiles he wrote *The Birds of Costa Rica* (1989), one of the first field guides to a tropical country. He chronicled his life in detail in books like *The Imperative Call* (1993), about his early adventures in Maryland, Jamaica,

Trogon collaris,
Collared Trogon

The Collared Trogon was one of the spectacular birds that convinced Skutch to abandon his study of bananas and concentrate on birds.

"For a large and growing number of people, birds are the strongest bond with the living world of nature. They charm us with lovely plumage and melodious songs; our quest of them takes us to the fairest places; to find them and uncover some of their well-guarded secrets we exert ourselves greatly and live intensely."

—Alexander F. Skutch (from "The Appreciative Mind," the epilogue to
A Bird Watcher's Adventures in Tropical America)

and Guatemala, and *A Naturalist in Costa Rica* (1971), perhaps his most-read book. Other subjects included his views on evolution (*Life Ascending*, 1985), and on religion (*The Quest of the Divine*, 1956). His last book, *Harmony and Conflict in the Living World* (2000), advocating a more peaceful co-existence with wildlife, was influenced by changes he witnessed around his forest home as industrial farming developed.

In addition to his many books, Skutch made innumerable contributions to journals and magazines, publishing nearly 200 papers. Roger Tory Peterson believed that Skutch's detailed life histories of Central American birds did for neotropical birds what Audubon's paintings did for the birds of North America.

He disliked statistics, preferring close observation and interpretation for his bird studies. Believing that even banding or ringing birds was wrong, he recognized individual birds on his estate by small differences in their plumage and behavior. Skutch thought that the minds of birds were revealed in the way they lived, behaviors that showed them to be "not unfeeling automata, but sensitive creatures aware of what they do."

His important discovery of "cooperative breeding" in Brown Jays, *Psilorhinus morio*, led to a lifelong interest in the ways birds help one another, especially in parenting and nest-building. He showed a clear preference for birds that got along with other birds, occasionally shooting at hawks when they threatened his preferred species. In 1987 he published *Helpers at Birds' Nests: A Worldwide Survey of Cooperative Breeding and Related Behavior*.

The Pamela and Alexander Skutch Research Award for studies in avian natural history was established in 1997 by the Association of Field Ornithologists and funded by an endowment from Skutch himself. In 2004, a few days before he died at 99, Skutch received the Loye and Alden Miller Research Award from the Cooper Ornithological Society for lifetime achievement in ornithological research.

Chamaepetes unicolor,
Black Guan

Visitors to the Alexander Skutch Los Cusingos Bird Sanctuary, situated on a property purchased by him in 1941, might see a Black Guan.

Psilorhinus morio,
Brown Jay

The Brown Jay of Central America has two color phases: In the north part of their range they are dark brown on top and lighter underneath, while in the south the population has a white belly.

R

Rabori *ra-BOR-eye*
After Dioscoro Rabor, a Filipino ornithologist, as in *Napothera* (now *Robsonius*) *rabori*, Cordillera Ground Warbler

Radiceus *ra-DIS-ee-us*
Rayed or striped, as in *Carpococcyx radiceus*, the Bornean Ground Cuckoo, with stripes on its underside

Rafflesii *RAF-fulz-ee-eye*
After Thomas Raffles, Lieutenant Governor of Java, as in *Dinopium rafflesii*, the Olive-backed Woodpecker

Raimondii *rye-MOND-ee-eye*
After Antonio Raimondi, an Italian-born Peruvian geographer and scientist, as in *Sicalis raimondii*, Raimondi's Yellow Finch

Rallicula *ral-li-KOO-la*
Rale, rail, and *-culus*, diminutive, as in *Rallicula rubra*, the Chestnut Forest Rail

Rallina *ral-LEEN-a*
Rale, rail, and diminutive, *-ina*, as in *Rallina rubra*, the Chestnut Forest Crake

Rallus *RAL-lus*
Rale, rail, and diminutive, *-ina*, as in *Rallina tricolor*, the Red-necked Crake

Ramphastos *ram-FASS-tos*
Greek, *rhamphos*, beak, and *-astus*, augmentative, as in *Ramphastos sulfuratus*, the Keel-billed Toucan

Ramphocaenus *ram-fo-SEE-nus*
Greek, *rhamphos*, beak, and *caen-*, new, fresh, as in *Ramphocaenus melanurus*, the Long-billed Gnatwren

Ramphocelus *ram-fo-SEL-us*
Greek, *rhamphos*, beak, and *kelas*, spot, as in *Ramphocelus nigrogularis*, the Masked Crimson Tanager

Ramphocinclus *ram-fo-SINK-lus*
Greek, *rhamphos*, beak, and *cinclus*, thrush, as in *Ramphocinclus brachyurus*, the White-breasted Thrasher

Ramphocoris *ram-fo-KOR-is*
Greek, *rhamphos*, beak, and *corys*, helmet, as in *Ramphocoris clotbey*, the Thick-billed Lark

Ramphodon *ram-FO-don*
Greek, *rhamphos*, beak, and *odon*, tooth, as in *Ramphodon naevius*, the Saw-billed Hermit

Ramphomicron *ram-fo-MY-kron*
Greek, *rhamphos*, beak, and *mikron*, small, as in *Ramphomicron dorsale*, the Black-backed Thornbill

Ramphotrigon *ram-fo-TRY-gon*
Greek, *rhamphos*, beak, and *trigon*, triangle, as in *Ramphotrigon megacephalum*, the Large-headed Flatbill, with the typical triangular bill of flycatchers

Ramsayi *RAM-zee-eye*
After Robert Ramsay, a British ornithologist, as in *Actinodura ramsayi*, the Spectacled Barwing

Randi *RAND-eye*
After Austen Rand, American ornithologist, as in *Muscicapa randi*, the Ashy-breasted Flycatcher

Randia *RAND-ee-a*
After Austen Rand, American ornithologist, as in *Randia pseudozosterops*, Rand's Warbler

Raphus *RAY-fus*
Raphus was the name assigned to bustards in 1760. Galenus, a Roman physician, named the Dodo *Raphus cucullatus*, referring to a bustard

Ramphotrigon megacephalum, Large-headed Flatbill

Rara *RAR-a*
Rare, as in *Phytotoma rara*, the Rufous-tailed Plantcutter

Rectirostris *rek-ti-ROSS-tris*
Recti-, straight, and *rostra*, bill, as in *Limnoctites rectirostris*, the Straight-billed Reedhaunter

Recurvirostra, -is *re-kur-vi-ROSS-tra/tris*
Recurvus, bent back, and *rostra*, bill, as in *Recurvirostra andina*, the Andean Avocet with an upcurved bill (see box)

Redivivum *re-di-VEE-um*
Revived, as in *Toxostoma redivivum*, the California Thrasher. Name refers to the fact that the bird was described, then "lost" and found again by another ornithologist

Reevei *REEVE-eye*
After J. P. Reeve, an American collector, as in *Turdus reevei*, the Plumbeous-backed Thrush

Reevesii *REEV-zee-eye*
After John Reeves, English naturalist and collector, as in *Syrmaticus reevesii*, Reeves's Pheasant

Regalis *re-GAL-is*
Regal, king, as in *Buteo regalis*, the Ferruginous Hawk

Regia, -us *RE-jee-a/us*
Royal, as in *Vidua regia*, the Shaft-tailed Whydah, probably because of the regal appearance of the male with a black crown and very long tail feathers

Regulorum *re-goo-LOR-um*
Royal, kingly, as in *Balearica regulorum*, the Gray Crowned Crane

Regulus, -oides *re-GOO-lus/re-goo-LOY-deez*
Diminutive of *rex*, little king or prince, as in *Regulus regulus*, the Goldcrest

Reichardi *RYE-cart-eye*
After Paul Reichard, German geographer and engineer, as in *Crithagra reichardi*, Reichard's Seedeater

Reichenbachii *RIKE-en-bak-ee-eye*
After Henrich Reichenbach, a German zoologist and botanist, as in *Anabathmis reichenbachii*, Reichenbach's Sunbird

Reichenowi *RIKE-ken-oh-eye*
After Anton Reichenow, German ornithologist, as in *Streptopelia reichenowi*, the White-winged Collared Dove

---

LATIN IN ACTION

Shorebird species have a variety of bill shapes and lengths so that they can exploit different food sources in their shoreline habitat. The recurved bill of the avocet is long and curves upward along its distal half. To catch its insect and invertebrate prey, the avocet sweeps its bill from side to side on the surface of the water. Avocet may come from the black and white outfits worn by European lawyers but the real etymology is unclear.

Recurvirostra andina, Andean Avocet

Reinwardtii *rine-VART-ee-eye*
After Caspar Reinwardt, Dutch ornithologist, as in *Apalharpactes reinwardtii*, the Javan Trogon

Reiseri *RYE-zer-eye*
After Othmar Reiser, Australian collector, as in *Phyllomyias reiseri*, Reiser's Tyrannulet

Religiosa *re-li-jee-OS-a*
Religious, sacred, revered, as in *Gracula religiosa*, the Common Hill Myna, which helps to propagate the Banyan Tree, sacred to Hindus

Remiz *RE-miz*
Polish for tit, as in *Remiz pendulinus*, the Eurasian Penduline Tit

Reticulata *re-ti-koo-LAT-a*
Reticulated, covered with ridges or grooves, as in *Meliphaga reticulata*, the Streak-breasted Honeyeater

Rex REKS
King, as in *Balaeniceps rex*, the Shoebill

Rhabdornis rab-DOR-nis
Greek, *rhabdotos*, striped, as in *Rhabdornis mystacalis*, the Stripe-headed Rhabdornis

Rhagologus rag-o-LO-gus
Greek, *rhago*, grape, berry, and *logos*, picked, chosen, as in *Rhagologus leucostigma*, the Mottled Whistler, which eats berries

Rhamphomantis ram-fo-MAN-tis
Greek, *rhamphos*, bill, and *mantis*, soothsayer, as in *Rhamphomantis* (now *Chrysococcyx*) *megarhynchus*, the Long-billed Cuckoo

Rhea REE-a
After a mythological Greek figure Rhea, the daughter of the sky god Uranus, as in *Rhea americana*, the Greater Rhea

Rheinardia rine-AR-dee-a
After Pierre-Paul Rheinhard, French Army officer, as in *Rheinardia ocellata*, the Crested Argus

Rhinocrypta rine-o-KRIP-ta
Greek, *rhinos*, nose, and *crypta*, hidden, as in *Rhinocrypta lanceolata*, the Crested Gallito; the nostrils are hidden by a bill plate

Rhodinocichla rosea,
Rosy Thrush-Tanager

Rhinomyias rine-o-MY-ee-as
Greek, *rhinos*, nose, and *muia*, fly, as in *Rhinomyias insignis*, the White-browed Jungle Flycatcher

Rhinopomastus rine-o-po-MAS-tus
Greek, *rhinos*, nose, and *pomos*, cover, as in *Rhinopomastus minor*, the Abyssinian Scimitarbill

Rhinoptilus rine-op-TIL-us
Greek, *rhinos*, nose, and *ptilon*, feather, as in *Rhinoptilus africanus*, the Double-banded Courser

Rhipidura, -us rip-ih-DOO-ra/rus
Greek, *rhipis*, a fan, and *oura*, tail, as in *Rhipidura nebulosa*, the Samoan Fantail

Rhizothera rise-o-THER-a
Greek, *rhiza*, root, and *thera*, hunting, pursuit, as in *Rhizothera longirostris*, the Long-billed Partridge

Rhodacanthis ro-da-KAN-thiss
Greek, *rhodon*, rose, and *akanthis*, finch, goldfinch, as in *Rhodacanthis flaviceps*, the extinct Lesser Koa Finch

Rhodinocichla ro-di-no-SIK-la
Greek, *rhodon*, rose, and *cichla*, thrush, as in *Rhodinocichla rosea*, the Rosy Thrush-Tanager

Rhodonessa ro-doe-NES-sa
Greek, *rhodon*, rose, and *nessa*, duck, as in *Rhodonessa caryophyllacea*, the probably extinct Pink-headed Duck

Rhodopechys ro-doe-PEK-is
Greek, *rhodon*, rose, and *pechys*, forearm, as in *Rhodopechys sanguineus*, the Eurasian Crimson-winged Finch

Rhinoptilus africanus,
Double-banded Courser

Rhodophoneus *ro-doe-FONE-ee-us*
Greek, *rhodon*, rose, and *phoneus*, a murderer, as in *Rhodophoneus* (now *Telophorus*) *cruentus*, the Rosy-patched Bushshrike

Rhodospiza *ro-doe-SPY-za*
Greek, *rhodon*, rose, and *spiza*, finch, as in *Rhodospiza obsoleta*, the Desert Finch, with pink patches on the wings

Rhodostethia *ro-doe-STETH-ee-a*
Greek, *rhodon*, rose, and *stethos*, breast, as in *Rhodostethia rosea*, Ross's Gull, with a rosy wash to the underparts, after British Rear Admiral James Ross who discovered the Ross Sea and Ross Ice Shelf

Rhopocichla *ro-po-SIK-la*
Greek, *rhopo*, bushes, brush, and *cichla*, thrush, as in *Rhopocichla atriceps*, the Dark-fronted Babbler

Rhopophilus *ro-po-FIL-us*
Greek, *rhopo*, bushes, brush, and *philos*, love, like, as in *Rhopophilus pekinensis*, the Chinese Hill Warbler

Rhopornis *ro-POR-nis*
Greek, *rhopo*, bushes, brush, and *ornis*, bird, as in *Rhopornis ardesiacus*, the Slender Antbird

Rhyacornis *ry-a-KOR-nis*
Greek, *rhya*, stream, and *ornis*, bird, as in *Rhyacornis bicolor*, the Luzon Water Redstart

Rhynchophanes *rin-ko-FAN-eez*
Greek, *rhynchos*, beak, and *phaino*, to appear, as in *Rhynchophanes mccownii*, McCown's Longspur

Rhynchopsitta *rin-kop-SIT-ta*
Greek, *rhynchos*, beak, and *psitta*, parrot, as in *Rhynchopsitta pachyrhyncha*, the Thick-billed Parrot

Rhynchortyx *rin-KOR-tiks*
Greek, *rhynchos*, beak, and *ortyx*, quail, as in *Rhynchortyx cinctus*, the Tawny-faced Quail

Rhynchotus *rin-KO-tus*
Greek, *rhunkhos*, beak, and *otus*, ear, as in *Rhynchotus rufescens*, the Red-winged Tinamou

Rhynochetos *rine-o-KET-os*
Greek, *rhinos*, nose, and *chetos*, corn (referring to corn-shaped flaps over the nostrils) as in *Rhynochetos jubatus*, the Kagu, a local native name

Rhopophilus pekinensis, Chinese Hill Warbler

Richardi *rich-ARD-eye*
After Richard of Luneville, French naturalist and collector, as in *Anthus richardi*, Richard's Pipit

Richardsii *RICH-ards-ee-eye*
After George Richards, British Rear-Admiral and geographer, as in *Ptilinopus richardsii*, the Silver-capped Fruit Dove

Ridgwayi *RIJ-way-eye*
After Robert Ridgway, American zoologist and curator, as in *Antrostomus ridgwayi*, the Buff-collared Nightjar

Ridibundus *ri-di-BUN-dus*
Ridere, to laugh, as in *Chroicocephalus ridibundus*, the Black-headed Gull, after the bird's call

Riparia *ri-PAR-ee-a*
Ripa, stream bank, as in *Riparia cincta*, the Banded Martin, which nests in sandbanks

Risoria *ri-SOR-ee-a*
Risor, one who mocks, as in *Streptopelia risoria* (now *roseogrisea*), the Barbary or African Collared Dove, referring to the bird's call

Rissa *RIS-sa*
From Islandic *rita*, kittiwake, as in *Rissa tridactyla*, the Black-legged Kittiwake

Robertsi *ROB-erts-eye*
After J. Austin Roberts, South African zoologist, as in *Oreophilais robertsi*, the Roberts's or Briar Warbler

Robinsoni *ro-bin-SON-eye*
After Herbert Robinson, British ornithologist and zoologist, as in *Myophonus robinsoni*, the Malayan Whistling Thrush

Robusta, -us *ro-BUST-a/us*
Robustus, of oak, hard, firm, as in *Gracula robusta*, the Nias Hill Myna, a stocky bird

Roraimae, -ia *ro-RIME-ee/ee-a*
After Mt. Roraima, Guyana and Venezuela, as in *Megascops roraimae*, the Roraiman Screech Owl

Rosea, -ata, -tus *rose-EE-a/rose-ee-AH-ta/tus*
Roseus, rose-colored, as in *Rhodostethia rosea*, Ross's Gull, with a pink wash to the underparts

Roseicapilla *rose-ee-eye-ka-PIL-la*
Roseus, rose-colored, and *capilla*, hair, as in *Ptilinopus roseicapilla*, the Mariana Fruit Dove

Roseicollis *rose-ee-eye-KOL-lis*
Roseus, rose-colored, and *colli-*, neck, as in *Agapornis roseicollis*, the Rosy-faced Lovebird

Roseifrons *rose-ee-EYE-fronz*
Roseus, rose-colored, and *frons*, forehead, as in *Pyrrhura roseifrons*, the Rose-fronted Parakeet

Roseigaster *rose-ee-eye-GAS-ter*
Roseus, rose-colored, and *gaster*, belly, as in *Priotelus roseigaster*, the Hispaniolan Trogon

Rosenbergi *RO-sen-berg-eye*
After Carl von Rosenberg, German naturalist and geographer, as in *Tyto rosenbergi*, the Sulawesi Masked Owl

Piranga roseogularis,
Rose-throated Tanager

Roseogrisea *rose-ee-a-GRISS-ee-a*
Roseus, rose-colored, and *grise*, gray, as in *Streptopelia roseogrisea*, the African Collared Dove

Roseogularis *rose-ee-o-goo-LAR-is*
Roseus, rose-colored, and *gula*, throat, as in *Piranga roseogularis*, the Rose-throated Tanager

Roseus *RO-zee-us*
Rose-colored, as in *Pastor roseus*, the Rosy Starling

Rossii *ROSS-ee-eye*
After Bernard Ross, Irish trader and administrator, as in *Chen rossii*, Ross's Goose

Rostratula, -us *ros-tra-TOO-la/lus*
Rostrum, beak, and *-atus*, with, as in *Rostratula australis*, the Australian Painted-snipe

Rostrhamus *ros-ter-HAM-us*
Rostrum, beak, and *hamus*, hook, as in *Rostrhamus sociabilis*, the Snail Kite, with a hooked beak

Rothschildi *ROTHS-child-eye*
After Lionel Walter Rothschild, founder of the Natural History Museum, Tring, England, as in *Leucopsar rothschildi*, the Bali Myna

Rubecula *roo-be-KOO-la*
Rubi, red, reddish, as in *Erithacus rubecula*, the European Robin

Ruber *ROO-ber*
Rubi, red, reddish, as in *Eudocimus ruber*, the Scarlet Ibis

Rubescens *roo-BES-sens*
Rubi, red, reddish, going to red, as in *Agraphospiza rubescens*, Blanford's Rosefinch or the Crimson Rosefinch

Rubiginosus *roo-bi-ji-NO-sus*
Reddish or rusty, as in *Colaptes rubiginosus*, the Golden-olive Woodpecker

Rubinus *roo-BYE-nus*
Rubi, red, reddish, as in *Pyrocephalus rubinus*, the Vermilion Flycatcher

Rubra *ROO-bra*
Rubi, red, reddish, as in *Paradisaea rubra*, the Red Bird-of-paradise

Rubricauda *roo-bri-KAW-da*
Rubi, red, reddish, and *cauda*, tail, as in *Phaethon rubricauda*, the Red-tailed Tropicbird

Rubriceps *ROO-bri-seps*
Rubi, red, reddish, and *ceps*, head, as in *Anaplectes rubriceps*, the Red-headed Weaver (see box)

Rubricollis *roo-bri-KOL-lis*
Rubi, red, reddish, and *collis*, neck, as in *Malimbus rubricollis*, the Red-headed Malimbe

Rubrifrons *ROO-bri-fronz*
Rubi, red, reddish, and *frons*, front, forehead, as in *Cardellina rubrifrons*, the Red-faced Warbler

Rubripes *roo-BRI-peez*
Rubi, red, reddish, and *pes*, foot, as in *Anas rubripes*, the American Black Duck

Rueppeli *roo-PEL-eye*
After Wilhelm Rüppell, a German collector, as in *Sylvia ruppeli*, Rüppell's Warbler

Rufa *ROO-fa*
Red, rufous, as in *Lessonia rufa*, the Austral Negrito

Rufescens *roo-FES-sens*
Reddish, as in *Atrichornis rufescens*, the Rufous Scrubbird

LATIN IN ACTION

Anaplectes comes from *Anapleko*, Greek for weave or braid, and lucidly describes the weaver birds or weaver finches of the family Plocidae, from the Greek *ploke*, a twining or weaving. And weave they do, constructing the most complex nests of any birds. Found mainly in sub-saharan Africa, the size, shape, and construction materials of the nests varies from species to species. The Red-headed Weaver (*Anaplectes rubriceps*) inhabits a wide area in southeastern Africa and exhibits a variety of plumage patterns, which have prompted ornithologists to assign it different scientific names over the years.

It was originally designated *Ploceus melanotis* (Black-eared Weaver) in 1839 even though some populations do not have black ear markings. In 1845 it became *Ploceus erythrocephalus* (Red-headed Weaver) but not until 1954 did *Anaplectes rubriceps* become the accepted name. Recent DNA evidence seems to indicate, however, the Red-headed Weaver is related to the *Ploceus* genus and should be *Ploceus rubriceps*. After almost 200 years, the name of this species is still being rethought.

Anaplectes rubriceps,
Red-headed Weaver

Ruficapilla, -lus *roo-fi-ka-PIL-la/lus*
Rufus, brownish-red, and *capilla*, hair, as in *Grallaria ruficapilla*, the Chestnut-crowned Antpitta

Ruficauda, -us, -atum *roo-fi-KAW-da/dus/ roo-fi-kaw-DAT-um*
Rufus, brownish-red, and *cauda*, tail, as in *Galbula ruficauda*, the Rufous-tailed Jacamar

Ruficeps *ROO-fi-seps*
Rufus, brownish-red and *ceps*, head, as in *Aimophila ruficeps*, the Rufous-crowned Sparrow

Ruficollis *roo-fi-COL-lis*
Rufus, brownish-red and *collis*, collar, neck, as in *Tachybaptus ruficollis*, the Little Grebe

Rufifrons *ROO-fi-fronz*
Rufus, brownish-red, and *frons*, front, forehead, as in *Formicarius rufifrons*, the Rufous-fronted Antthrush

Rufigula, -aris *roo-fi-GOO-la/roo-fi-goo-LAR-is*
Rufus, brownish-red, and *gula*, throat, as in *Ficedula rufigula*, the Rufous-throated Flycatcher

Rufinucha *roo-fi-NOO-ka*
Rufus, brownish-red, and *nucha*, nape, as in *Atlapetes rufinucha*, the Bolivian Brush Finch

Rufipectus *roo-fi-PEK-tus*
Rufus, brownish-red, and *pectus*, breast, as in *Leptopogon rufipectus*, the Rufous-breasted Flycatcher

Rufipennis *roo-fi-PEN-nis*
Rufus, brownish-red, and *pennis*, feather, as in *Butastur rufipennis*, the Grasshopper Buzzard

Rufiventer, -tris *roo-fi-VEN-ter/tris*
Rufus, brownish-red, and *venter*, belly, abdomen, as in *Tachyphonus rufiventer*, the Yellow-crested Tanager

Rufivirgata, -us *roo-fi-vir-GAT-a/us*
Rufus, brownish-red, and *virgata*, striped, as in *Arremonops rufivirgatus*, the Olive Sparrow

Rufogularis *roo-fo-goo-LAR-is*
Rufus, brownish-red, and *gula*, throat, as in *Alcippe rufogularis*, the Rufous-throated Fulvetta

Rufum, -us *ROO-fum/fus*
Rufus, brownish-red, as in *Toxostoma rufum*, the Brown Thrasher

Rupestris *roo-PES-triss*
Rock dweller, as in *Columba rupestris*, the Hill Pigeon, which nests on cliffs and rock ledges

Rupicola *roo-pi-KO-la*
Rupes, cliff, and *cola*, inhabitant, as in *Rupicola peruvianus*, the Andean Cock-of-the-rock

Rustica *RUSS-ti-ka*
Rusticus, rural, country, as in *Hirundo rustica*, the Barn Swallow, which avoids urban areas

Rusticola, -us *rus-ti-KOL-a/us*
Rusticus, rural, country, and *cola*, inhabitant, as in *Scolopax rusticola*, the Eurasian Woodcock

Ruticilla *roo-ti-SIL-la*
Rutilis, reddish, and *cilla*, tail, as in *Setophaga ruticilla*, the American Redstart

Rynchops *RIN-kops*
Greek, *rynchas*, beak, and *ops*, face, as in *Rynchops niger*, the Black Skimmer

Rupicola peruvianus sanguinolentus, Andean Cock-of-the-rock (subspecies)

S

Sabini *SAY-bine-eye*
After Joseph Sabine, English zoologist, as in *Rhaphidura sabini*, Sabine's Spinetail

Sagittarius *sa-jit-TAR-ee-us*
An archer, as in *Sagittarius serpentarius*, the Secretarybird; may refer to the feathers that remind one of an archer's quiver of arrows or the way the bird walks, as an archer stalking its prey

Salmoni *SAL-mon-eye*
After Thomas Salmon, Colombian engineer, as in *Brachygalba salmoni*, the Dusky-backed Jacamar

Salpinctes *sal-PINK-teez*
Salpinx, trumpet, as in *Salpinctes obsoletus*, the Rock Wren; Ancient Greeks compared the song of the Winter Wren to a trumpet and the genus name became applied to the North American Rock Wren

Salpornis *sal-POR-nis*
Greek, *salpinx*, trumpet, and *ornis*, bird, as in *Salpornis spilonotus*, the Indian Spotted Creeper, with a high-pitched call

Saltator *sal-TAY-tor*
Leaper, dancer, as in *Saltator maximus*, the Buff-throated Saltator; scientific and common name derive from the heavy hopping of the birds on the ground

Salvadorii, -ia *sal-va-DOR-ee-eye/ee-a*
After Conte Salvadori, Italian physician, educator, and ornithologist, as in *Cryptospiza salvadorii*, the Abyssinian Crimsonwing

Salvini *SAL-vin-eye*
After Osbert Salvin, English naturalist, as in *Antrostomus salvini*, the Tawny-collared Nightjar

Samarensis *sam-a-REN-sis*
After Samar, Phillipines, as in *Orthotomus samarensis*, the Yellow-breasted Tailorbird

Samoensis *sam-o-EN-sis*
After Samoa, as in *Zosterops samoensis*, the Samoan White-eye

Antrostomus salvini,
Tawny-collared Nightjar

Sanctithomae *sank-ti-TO-mee*
After São Tomé, as in *Ploceus sanctithomae*, the São Tomé Weaver

Sandwichensis, -vicensis *sand-wich-EN-sis/sand-vi-SEN-sis*
After the Sandwich Islands (Hawaii) as in *Porzana sandwichensis*, the extinct Hawaiian Rail

Sanfordi *SAN-ford-eye*
After Leyland Sanford, American zoologist, as in *Cyornis sanfordi*, the Matinan Blue Flycatcher

Sanguinea, -us *san-GWIN-ee-a/us*
Sangui, blood, as in *Cacatua sanguinea*, the Little Corella, with dark pink markings around the bill and in front of the eyes

Sanguiniceps *san-GWIN-ih-seps*
Sangui, blood, and *ceps*, head, as in *Haematortyx sanguiniceps*, the Crimson-headed Partridge. Scientific name literally means bloody quail, bloody head

Sanguinodorsalis *san-gwin-oh-dor-SAL-is*
Sangui, blood, and *dorsum*, back, as in *Lagonosticta sanguinodorsalis*, the Rock Firefinch

Sarcogyps *SAR-ko-jips*
Greek, *sarc*, flesh, and *gyps*, vulture, as in *Sarcogyps calvus*, the Red-headed Vulture

Sarcops *SAR-kops*
Greek, *sarc*, flesh, and *ops*, face, appearance, as in *Sarcops calvus*, the Coleto, with an unfeathered head

Sarcoramphus *sar-ko-RAM-fus*
Greek, *sarc*, flesh, and *ramphos*, beak, as in *Sarcoramphus papa*, the King Vulture

Sarkidiornis *sar-kid-ee-OR-nis*
Greek, *sarc*, flesh, *idios*, distinct, and *ornis*, bird, as in *Sarkidiornis melanotos*, the Knob-billed Duck

Saroglossa *sar-o-GLOSS-a*
Greek, *saro*, broom, and *glossa*, tongue, as in *Saroglossa* (now *Hartlaubius*) *auratus*, the Madagascan Starling, which has small brushlike spines on its tongue

Sarothrura *sar-oth-RUR-a*
Greek, *saro*, broom, and *oura*, tail, as in *Sarothrura ayresi*, the White-winged Flufftail

Saturata, -us *sa-tur-AT-a/us*
Saturated, as with color, as in *Euphonia saturata*, the Orange-crowned Euphonia

Saundersi *SAWN-ders-eye*
After Howard Saunders, British ornithologist, as in *Chroicocephalus saundersi*, Saunders's Gull

Saurophagus *sore-o-FAY-gus*
Greek, *sauro*, lizard, and *phagein*, eat, as in *Todiramphus saurophagus*, the Beach Kingfisher; its diet includes lizards

Savilei *sa-VIL-eye*
After Robert Savile, British soldier and diplomat, as in *Lophotis savilei*, Savile's Bustard

Sawtelli *SAW-tel-lye*
After Gorden Sawtell, British civil servant, as in *Aerodramus sawtelli*, the Atiu Swiftlet

Saxicola, -lina, -oides *saks-ih-KO-la/saks-ih-ko-LEEN-a/saks-ih-ko-LOY-deez*
Saxum, a stone, *colo*, inhabit, as in *Saxicola rubetra*, the Whinchat, common in open rock-strewn habitats

Sayornis and Saya *say-OR-nis and SAY-a*
After Thomas Say, American naturalist and zoologist, as in *Sayornis saya*, Say's Phoebe

Scandens, -iacus *SKAN-denz/skan-dee-AK-us*
Scand-, climbing, as in *Geospiza scandens*, the Common Cactus Finch; probably because they climb around cactus plants to feed on the nectar in their flowers

Scardafella *skar-da-FEL-la*
From Dante, to express scaliness, as in *Scardafella* (now *Columbina*) *inca*, the Inca Dove

Scelorchilus *skel-or-KIL-us*
Greek, *skelos*, leg, and *orkhilos*, wren, as in *Scelorchilus rubecula*, the Chucao Tapaculo, which resembles a wren

Scenopoeetes *sken-o-po-EE-teez*
Greek, *skene*, a covered place, and *poietes*, maker, as in *Scenopoeetes dentirostris*, the Tooth-billed Bowerbird, which makes a covered bower

Schalowi *SHAL-o-eye*
After Herman Schalow, German banker, as in *Tauraco schalowi*, Schalow's Turaco

Scheepmakeri *SHEP-mak-er-eye*
After C. Scheepmaker, Dutch civil servant and collector, as in *Goura scheepmakeri*, the Southern Crowned Pigeon

Schistacea, -us *shis-TAY-see-a/us*
Schistus, slate, as in *Sporophila schistacea*, the Slate-colored Seedeater

Schisticeps *SHIS-ti-seps*
Schistus, slate, and *-ceps*, head, as in *Coracina schisticeps*, the Gray-headed Cuckooshrike

Geospiza scandens, Common Cactus Finch

Schistochlamys shis-to-KLAM-is
Greek, *schistus*, slate, and *khlamus*, cloak, as in *Schistochlamys ruficapillus*, the Cinnamon Tanager

Schlegelii shlay-GEL-ee-eye
After Hermann Schlegel, German zoologist, as in *Pachycephala schlegelii*, the Regent Whistler

Schneideri SHNYE-der-eye
After Gustav Schneider, Swiss zoologist, as in *Hydrornis schneideri*, Schneider's Pitta

Scissirostrum shis-si-ROSS-trum
Scissi, cut, split, and *rostrum*, beak, as in *Scissirostrum dubium*, the Grosbeak Starling, with a powerful bill

Sclateri, -a SKLAY-ter-eye/a
After Philip or William Sclater, British naturalist, as in *Nonnula sclateri*, the Fulvous-chinned Nunlet

Sclerurus skler-OO-rus
Greek *skler*, hard, and *oura*, tail, as in *Sclerurus rufigularis*, the Short-billed Leaftosser, with a stiff tail

Scolopaceus sko-lo-PACE-ee-us
Greek, *skolopax*, woodcock, as in *Limnodromus scolopaceus*, the Long-billed Dowitcher, a bird like a woodcock or snipe

Scolopax SKO-lo-paks
Greek, *skolopax*, woodcock, as in *Scolopax saturata*, the Javan Woodcock

Scopus SKO-pus
Scopae, a broom of twigs, as in *Scopus umbretta*, the Hamerkop, which builds a huge nest of twigs up to 1.5 meters across

Scotocerca sko-toe-SIR-ka
Greek, *scotos*, darkness, and *cercos*, tail, as in *Scotocerca inquieta*, the Streaked Scrub Warbler, with a tail darker than the rest of its body

Scotopelia sko-toe-PEL-ee-a
Greek, *scotos*, darkness, and *peleia*, pigeon, as in *Scotopelia bouvieri*, the Vermiculated Fishing Owl (see box); no explanation of pigeon vs owl

Scutatus, -a skoo-TAT-us/a
Scutum, shield, as in *Malimbus scutatus*, the Red-vented Malimbe, which has a bright-red upper chest and throat resembling a shield

---

LATIN IN ACTION

The Vermiculated Fishing Owl (*Scotopelia bouvieri*) is described as a nocturnal dove by its genus. Nocturnal is appropriate, but dove does not appear to be. Vermiculated, which means worm-like, as in wavy lines, might not be the best descriptor either. The breast of this bird is streaked with heavy vertical lines; the back and wings have much more muted but still wavy markings. Hunting along river edges in central Africa, it swoops down on fish, frogs, and insects on the river surface, rarely entering the water even partially. They feed by sight as the sound of the river would mask the subtle sounds of their prey.

Scotopelia bouvieri, Vermiculated Fishing Owl

Scytalopus skit-a-LOP-us
Greek, *scutale*, thick stick, and *pous*, foot, as in *Scytalopus latrans*, the Blackish Tapaculo; an allusion to the relatively heavy legs and feet

Seebohmi SEE-bome-eye
After Henry Seebohm, British businessman and amateur ornithologist, as in *Amphilais seebohmi*, the Gray Emutail

Seicercus sy-SIR-kus
Greek, *sei*, shake, and *cercos*, tail, as in *Seicercus grammiceps*, the Sunda Warbler; presumably named for its tail movements

Migration

A variety of animals migrate long or short distances from wintering to breeding grounds and back again, or they wander around in search of food, but birds have incorporated these annual movements into their life cycles like almost no other creature. Birds migrate mainly away from nesting sites as cold weather approaches and food supplies decline, and head to warmer wintering areas with abundant fruit, seeds, insects, and other food items. As spring approaches in their breeding grounds, they leave their winter home and migrate back there, now that they have sufficient food, potential mates, and nest sites.

Food and reproduction are the evolutionary (ultimate) reasons for moving to spring breeding grounds; food and escape from winter weather are the ultimate reasons for migrating to warmer wintering grounds. The timing of migration, though, is cued not by weather factors like temperature, but by genetic factors, hormone levels, and particularly photoperiod, the amount of daylight. As the days get longer, migratory birds on their wintering grounds show what is called "migratory restlessness" and eventually take off on their journey to their breeding grounds (north in the northern hemisphere, the opposite in the southern hemisphere). Conversely, as the days shorten in the breeding areas, the birds reverse the journey. Weather does have some effect on their flights, though. Low pressure with wind and rain may slow the speed of migration, and continued good weather may induce them to stay where they are, at least for a short time.

Instead of migrating across several lines of latitude, some birds simply move down from higher elevations. This is called altitudinal migration. White-ruffed Manakins (*Corapipo altera*) of Central America, for example, migrate to lower elevations during the rainy season to avoid heavy rainstorms.

Birds migrate along flyways (there are eight major ones worldwide) from their breeding to wintering grounds and back. Whether across North America to Central and South America or across Europe to Africa, migrating birds follow general routes that vary with the species and that have developed over evolutionary time to provide the most efficient path to their destination. The Arctic Tern (*Sterna paradisaea*), the animal with the longest migration, follows coastlines from the Arctic to the Antarctic and back again, covering about 44,000 miles (73,000 kilometers) each year.

Limosa lapponica,
Bar-tailed Godwit

A female Bar-tailed Godwit fitted with a satellite transmitter was found to have completed a 7,000-mile (11,500-kilometer) non-stop flight from Alaska to New Zealand.

MIGRATION

Sterna paradisaea,
Arctic Tern

The Arctic Tern has the longest migratory trip of any animal, from the Arctic to the Antarctic and back again every year, over 4,350 miles (70,000 kilometers).

Ornithologists have studied bird migration for over 2,000 years by putting numbered bands or rings on birds and noting where and when they were captured and recaptured. This is called "ringing" in the UK and Europe and "banding" in the US. Of birds that are hunted, like ducks and geese, about 16 percent of the bands are eventually retrieved. With songbirds it is less than 1 percent. More recent techniques involve the use of small transmitters to track birds by radio or microwave telemetry. Radar is also used to follow large flocks of birds from one area to another.

Data have been collected for many years on the dates of migratory arrival and departure each year, and it is clear that many birds have moved their migratory dates earlier due to global warming. Climate change causes flowers, insects, and seeds to appear earlier, and to be the best competitors for food and mates, birds have to arrive early as well. Although birds respond to photoperiod and not weather, there is always variability in a population and when conditions are changing the birds that migrate earlier become the most successful individuals.

To find their way from one part of the earth to another, birds need to have some mechanism of navigation. Birds can use landmarks such as lakes, rivers, and mountain ranges, to gauge their path. But landmarks are not the only way. A population of the Bar-tailed Godwit (*Limosa lapponica*) migrates from New Zealand to China, non-stop across the ocean, for 6,000 miles (10,000 kilometers) each year. Birds also use the position of the sun, the moon, and the stars. And recent evidence has shown that birds can detect geomagnetic lines of force via their ophthalmic nerves. Most birds probably use two or more of these mechanisms, but nevertheless they occasionally lose their way or are blown off course by a storm. A number of islands have been colonized this way. A flock of Fieldfares (*Turdus pilaris*) were pushed to Greenland by a storm and naturalized there.

Turdus pilaris,
Fieldfare

The Fieldfare, now a resident of southern Greenland, found its way to the country-island when a group of the birds was blown off course during a migratory flight.

Seiurus *see-eye-OO-rus*
Greek, *sei*, shake, and *oura*, tail, as in *Seiurus aurocapilla*, the Ovenbird. While walking it holds its tail high but while resting it slowly moves it up and down

Selasphorus *sel-as-FOR-us*
Greek, *selas*, light, and *phoros*, bearing, as in *Selasphorus flammula*, the brightly colored Volcano Hummingbird

Selenidera *sel-en-ih-DER-a*
Greek, *selene*, moon, and *dera*, neck, throat, as in *Selenidera maculirostris*, the Spot-billed Toucanet, with crescent-shaped spots on its bill

Seleucidis *sel-loy-SID-is*
Greek, *seleukidos*, a locust-eating bird, as in *Seleucidis melanoleucus*, the Twelve-wired Bird-of-paradise

Semicinerea, -us *se-mee-sin-AIR-ee-a/us*
Semi, half, and *ciner-*, ashy, as in *Cranioleuca semicinerea*, the Gray-headed Spinetail

Semicollaris *se-mee-col-LAR-is*
Semi, half, and *collaris*, neck, collar, as in *Nycticryphes semicollaris*, the South American Painted-snipe

Semifasciata *se-mee-fas-see-AT-a*
Semi, half, and *fasciat-*, banded, as in *Tityra semifasciata*, the Masked Tityra

Semifuscus *se-mee-FUS-kus*
Semi, half, and *fusc-*, dusky, as in *Chlorospingus semifuscus*, the Dusky Bush Tanager

Semipalmatus *se-mee-pal-MAT-us*
Semi, half, and *palmatus*, palm, as in *Charadrius semipalmatus*, the Semipalmated Plover, with partially webbed feet

Semiplumbeus *se-mee-PLUM-bee-us*
Semi, half, and *plumbeus*, lead (colored), as in *Rallus semiplumbeus*, the Bogota Rail

Semirufa, -us *se-mee-ROOF-a/us*
Semi, half, and *rufa*, rufus, as in *Cecropis semirufa*, the Red-breasted Swallow

Semitorquata, -us *se-mee-tor-KWAT-a/us*
Semi, half, and *torquatus*, collared, necklace, as in *Ficedula semitorquata*, the Semicollared Flycatcher

Semnornis *sem-NOR-nis*
Greek, *semnos*, fine, great, and *ornis*, bird, as in *Semnornis frantzii*, the Prong-billed Barbet

Senegala, -oides, -allus, -ensis *sen-eh-GAL-a/sen-eh-gal-OY-deez/sen-eh-GAL-lus/sen-eh-gal-EN-sis*
From Senegal, as in *Centropus senegalensis*, the Senegal Coucal

Sericornis *se-ri-KOR-nis*
Greek, *serikos*, silken, and *ornis*, bird, as in *Sericornis keri*, the Atherton Scrubwren; presumably from the silky-appearing plumage of their back and head feathers

Sericulus *se-ri-KOO-lus*
Greek, *serikos*, silken, and diminutive *-culus*, as in *Sericulus aureus*, the Masked Bowerbird, with silky plumage

Serinus *ser-EYE-nus*
Serinus, referring to a bird called the serin, as in *Serinus canaria*, the Atlantic Canary

Serpophaga *ser-po-FAY-ga*
Greek, *serphos*, small insect, and *phagein*, to eat, as in *Serpophaga hypoleuca*, the River Tyrannulet

Serrator *ser-RA-tor*
Serra, saw, as in *Mergus serrator*, the Red-breasted Merganser

Setophaga *se-toe-FAY-ga*
Greek, *setos*, insect, and *phagein*, to eat, as in *Setophaga citrina*, the Hooded Warbler

Rallus semiplumbeus, Bogota Rail

SPATULA

Sialia currucoides,
Mountain Bluebird

Sewerzowi *su-er-ZO-eye*
After Nikolai Severzov (sic), Russian zoologist, as in *Tetrastes sewerzowi*, the Chinese Grouse

Sharpei, -ii *SHARP-eye/ee-eye*
After Richard Sharpe, British zoologist, as in *Macronyx sharpei*, Sharpe's Longclaw

Shelleyi *SHEL-lee-eye*
After George Shelley, British geologist and ornithologist, as in *Nesocharis shelleyi*, Shelley's Oliveback

Sialia *see-AL-ee-a*
Greek, *sialis*, a word used by Aristotle to refer to an unidentified bird, as in *Sialia currucoides*, the Mountain Bluebird

Sibilatrix *si-bi-LA-tricks*
Sibila, whistle, as in *Phylloscopus sibilatrix*, the Wood Warbler

Sieboldii *see-BOLD-ee-eye*
After Philip von Siebold, German physician and naturalist, as in *Treron sieboldii*, the White-bellied Green Pigeon

Signatus *sig-NA-tus*
Signare, mark, stamp, designate, as in *Knipolegus signatus*, the Andean Tyrant; perhaps due to the distinctive wing whirring sound the male makes during courtship

Similis *si-MIL-is*
Like, as in *Anthus similis*, the Long-billed Pipit, with a number of similar-looking geographic races

Simplex *SIM-pleks*
Simple, as in *Piculus simplex*, the Rufous-winged Woodpecker

Sinaloa, -ae *sin-a-LOW-a/ee*
After Sinaloa, Mexico, as in *Thryophilus sinaloa*, the Sinaloa Wren

Sinensis *si-NEN-sis*
Referring to China, Chinese, as in *Sturnia sinensis*, the White-shouldered Starling

Sitta *SIT-ta*
Greek, *sitte*, a kind of woodpecker or prober, as in *Sitta castanea*, the Indian Nuthatch, which climbs trees like a woodpecker

Sittasomus *sit-ta-SO-mus*
Greek, *sitte*, a kind of woodpecker or prober, and *soma*, body as in *Sittasomus griseicapillus*, the Olivaceous Woodcreeper

Sittiparus *sit-ti-PAR-us*
Greek, *sitte*, a kind of woodpecker or prober, and *parus*, tit, as in *Sittiparus varius*, the Varied Tit, which probes for insects and seeds

Smicrornis *smik-ROR-nis*
Greek, *smikros*, small, and *ornis*, bird, as in *Smicrornis brevirostris*, the Weebill

Solitaria, -us, -ius *sol-ih-TAR-ee-a/ us/ee-us*
Solitary, as in *Tringa solitaria*, the Solitary Sandpiper, which tends not to be found in large groups

Somateria *so-ma-TAIR-ee-a*
Greek, *soma*, body, and *erion*, down, as in *Somateria mollissima*, the Common Eider; its soft body down is used in quilts

Sordida, -us, -ulus *sor-DI-da/dus/sor-di-DOO-lus*
Dirty, unkempt, as in *Pinarochroa sordida*, the Moorland Chat

Spatula *spat-OO-la*
Spoon, as in *Spatula* (now *Anas*) *clypeata*, the Northern Shoveler, with a wide and flat bill

Somateria mollissima,
Common Eider

Speciosa *spe-see-O-sa*
Species, beautiful, as in *Dasycrotapha speciosa*, the Flame-templed Babbler

Spectabilis *spek-TA-bil-is*
Fancy, showy, as in *Somateria spectabilis*, the King Eider

Speirops *SPY-rops*
Greek, *speira*, wound around, and *ops*, eye, as in *Speirops* (now *Zosterops*) *lugubris*, the Black-capped Speirops; name refers to the bird's white eye ring

Spelaeornis *spel-ee-OR-nis*
Greek, *speos*, cave, and *ornis*, bird, as in *Spelaeornis caudatus*, the Rufous-throated Wren-Babbler, which builds its nest in very thick brush, as if in a cave

Speotyto *spee-o-TI-to*
Greek, *speos*, cave, and *tyto*, owl, as in *Speotyto* (now *Athene*) *cunicularia*, the Burrowing Owl

Spermestes *sper-MESS-teez*
Greek, *sperma*, seed, and Latin *estes*, eating, as in *Spermestes* (now *Lonchura*) *cucullata*, the Bronze Mannikin

Spermophaga *sper-mo-FAY-ga*
Greek, *sperma*, seed, and *phagein*, eating, as in *Spermophaga haematina*, the Western Bluebill

Spheniscus *sfen-ISS-kus*
Greek, *sphen*, a wedge, and *-icus*, diminutive, as in *Spheniscus humboldti*, the Humboldt Penguin, after its flipper-like wings

Sphenocichla *sfen-o-SIK-la*
Greek, *sphen*, a wedge, and Latin, *cichla*, a thrush, as in *Sphenocichla roberti*, the Cachar Wedge-billed Babbler

Sphecotheres *sfee-ko-THER-eez*
Greek, *sphekos*, a wasp, and *therao*, hunt, as in *Sphecotheres viridis*, the Green Figbird, which eats insects and occasionally wasps

Sphyrapicus *spy-RAP-ih-kus*
Greek, *sphyra*, hammer, and Latin, *picus*, woodpecker, as in *Sphyrapicus nuchalis*, the Red-naped Sapsucker

Spilocephalus *spil-o-se-FAL-us*
Greek, *spilos*, spot, and Latin, *cephala*, head, as in *Otus spilocephalus*, the Mountain Scops Owl; the top of its head is spotted

Spilodera *spil-o-DARE-a*
Greek, *spilos*, spot, and *der*, neck, hide, as in *Petrochelidon spilodera*, the South African Cliff Swallow

Spilogaster *spil-o-GAS-ter*
Greek, *spilos*, spot, and *gaster*, belly, as in *Aquila spilogaster*, the African Hawk-Eagle

Spilonotus *spil-o-NO-tus*
Greek, *spilos*, spot, and *noto*, back, as in *Circus spilonotus*, the Eastern Marsh Harrier, with a spotted back

Spilornis *spil-OR-nis*
Greek, *spilos*, spot, and *ornis*, a bird, as in *Spilornis cheela*, the Crested Serpent Eagle, with a spotted underside

Spinus *SPINE-us*
Greek, *spinos*, linnet or siskin, as in *Spinus spinus*, the Eurasian Siskin

Spixii *SPIKS-ee-eye*
After Johann Von Spix, German naturalist, as in *Cyanopsitta spixii*, Spix's Macaw

Spiza *SPY-za*
Greek, *spiza*, finch, as in *Spiza americana*, the Dickcissel, common name from their call

Spizaetus *spy-ZEE-tus*
Greek, *spizias*, hawk, and *aetos*, eagle, as in *Spizaetus ornatus*, the Ornate Hawk-Eagle; birds of this genus are intermediate in size between hawks and eagles

Spizella *spy-ZEL-la*
Greek, *spiza*, finch, and Latin, *-ella*, diminutive, as in *Spizella passerina*, the Chipping Sparrow

Sporophila *spo-ro-FIL-a*
Greek, *sporos*, seed, and *philos*, loving, as in *Sporophila frontalis*, the Buffy-fronted Seedeater

Squamata, -tus *skwa-MA-ta/tus*
Squamatus, scaled, as in *Eos squamata*, the Violet-necked Lory

Spiza americana, Dickcissel

Squatarola skwa-ta-RO-la
A type of plover, as in *Pluvialis squatarola*, the Gray Plover

Stachyris sta-KIR-is
Greek, *stachus*, head of grain, and *rhis*, nose, as in *Stachyris grammiceps*, the White-breasted Babbler, referring to the flap of tissue nearly covering the nares (nostrils)

Steatornis stee-a-TOR-nis
Greek, *steatos*, fat, and *ornis*, bird, as in *Steatornis caripensis*, the Oilbird

Steerii STEER-ee-eye
After Joseph Steere, American ornithologist, as in *Sarcophanops steerii*, the Wattled Broadbill (see box)

Stelgidopteryx stel-ji-DOP-ter-iks
Greek, *stelgis*, scraper, and *pteryx*, wing, as in *Stelgidopteryx ruficollis*, the Southern Rough-winged Swallow

Stelleri STEL-ler-eye
After George Steller, German naturalist and explorer, as in *Polysticta stelleri*, Steller's Eider

Stephanoaetus ste-fan-o-EE-tus
Greek, *stephano*, crown, and *aetos*, eagle, as in *Stephanoaetus coronatus*, the Crowned Eagle

Stercorarius ster-ko-RARE-ee-us
Stercus, excrement, as in *Stercorarius parasiticus*, the Arctic Skua or Parasitic Jaeger, which pursues other birds to force them to regurgitate their food, the ejecta once thought to be excrement

Sterna STER-na
Latin form of the English tern, as in *Sterna hirundo*, the Common Tern

Stictonetta stik-toe-NET-ta
Greek, *stiktos*, spotted, dotted, and *netta*, duck, as in *Stictonetta naevosa*, the Freckled Duck

Stiphrornis stif-ROR-nis
Greek, *stiphros*, firm, and *ornis*, a bird, as in *Stiphrornis erythrothorax*, the Forest Robin, with a stocky build for an Old World Flycatcher

Stolzmanni STOLZ-man-nye
After Jan Sztolcmann, Polish ornithologist, as in *Urothraupis stolzmanni*, the Black-backed Bush Tanager

---

LATIN IN ACTION

The Wattled Broadbill, *Eurylaimus steerii*, restricted to a small area on Mindanao in the Philippines, behaves like a flycatcher, sallying out from a branch to catch an insect and beating the large ones against a branch before eating them. Like other flycatchers, its bill is wide and has a small hook at the tip, but the bill is much heavier than the flycatchers'.

Sarcophanops steerii, Wattled Broadbill

Strepera stre-PAIR-a
Streperus, noisy, as in *Strepera fuliginosa*, the Black Currawong, a loud and noisy bird

Streptopelia strep-to-PIL-ee-a
Strepto, twisted, and *peleia*, dove, as in *Streptopelia turtur*, the European Turtle Dove; refers to the markings around the neck

Stresemanni STREZ-man-nye
After Erwin Stresemann, German ornithologist and collector, as in *Merulaxis stresemanni*, Stresemann's Bristlefront

Striata, -us stree-AT-a/us
Striated, as in *Butorides striata*, the Striated Heron

Striaticeps stree-AT-ih-seps
Striata, striated, and *ceps*, head, as in *Knipolegus striaticeps*, the Cinereous Tyrant

Striaticollis stree-at-ih-KOL-lis
Striata, striated, and *collis*, collar, neck, as in *Fulvetta striaticollis*, the Chinese Fulvetta

Striatus *stree-AT-us*
Striata, striated, as in *Colius striatus*, the Speckled Mousebird

Stricklandii *strik-LAND-ee-eye*
After Hugh Strickland, British geologist and naturalist, as in *Gallinago stricklandii*, the Fuegian Snipe

Strigops *STRY-gops*
Greek, *strigos*, a night bird, and *ops*, eye, as in *Strigops habroptila*, the Kakapo, a Maori word for night parrot

Strix *STRIKS*
Greek, *strigx*, utter shrill sounds, as in *Strix ocellata*, the Mottled Wood Owl

Struthio *STROO-thee-o*
Struthio, ostrich, shortened from *struthiocamelus*, the camel sparrow, because of its size, as in *Struthio camelus*, the Common Ostrich

Sturnella *stir-NEL-la*
Diminutive of *sturnus*, starling, as in *Sturnella magna*, the Eastern Meadowlark

Sturnus *STIR-nus*
Starling, as in *Sturnus vulgaris*, the Common or European Starling

Subalaris *sub-a-LAR-is*
Sub, under, *ala*, wing, arm, as in *Turdus subalaris*, the Eastern Slaty Thrush

Strigops habroptila,
Kakapo

Subcristata *sub-kris-TA-ta*
Sub, under, *cristatus*, crested, as in *Cranioleuca subcristata*, the Crested Spinetail

Sula *SOO-la*
Icelandic, *sula*, gannet, as in *Sula nebouxii*, the Blue-footed Booby, booby from Spanish slang *bobo*, stupid

Superba, -us *soo-PERB-a/us*
Super, superb, as in *Cyornis superbus*, the Bornean Blue Flycatcher

Superciliaris *soo-per-sil-ee-AR-is*
Supercilium, eyebrow, as in *Camaroptera superciliaris*, the Yellow-browed Camaroptera

Superciliosa, -um, -us
soo-per-sil-ee-OS-a/um/us
Supercilium, eyebrow, as in *Poecile superciliosus*, the White-browed Tit

Swainsoni, -ii *SWAIN-son-eye/swain-SON-ee-eye*
After William Swainson, British naturalist and illustrator, as in *Buteo swainsoni*, Swainson's Hawk

Swinhoii *swin-HO-ee-eye*
After Robert Swinhoe, Indian naturalist and collector, as in *Lophura swinhoii*, Swinhoe's Pheasant

Swynnertoni, -ia *SWIN-ner-ton-eye/ee-a*
After Charles Swynnerton, Indian-born entomologist, as in *Swynnertonia swynnertoni*, Swynnerton's Robin

Sylvaticus *sil-VAT-ih-kus*
Silvaticus, of the woods, as in *Turnix sylvaticus*, the Common Buttonquail

Sylvia *SIL-vee-a*
Silva, a forest, as in *Sylvia borin*, the Garden Warbler, which despite its common name frequents dense undergrowth

Synallaxis *sin-al-LAK-sis*
From French *Synallaxe* for spinetails, as in *Synallaxis albigularis*, the Dark-breasted Spinetail

Synthliboramphus *sin-th-lih-bo-RAM-fus*
Synthlibo, to press, and *ramphus*, beak, as in *Synthliboramphus hypoleucus*, the Guadalupe Murrelet, referring to the laterally compressed beak

Syrmaticus *sir-MAT-ih-kus*
Greek, *syrma*, trailing robe, as in *Syrmaticus soemmerringii*, the Copper Pheasant

Lophura swinhoii,
Swinhoe's Pheasant

Margaret Morse Nice
(1883–1974)

Margaret Morse Nice was a major force in changing the way ornithologists looked at birds, from checking them off to collecting data on their behavior.

Margaret Morse Nice was an American ornithologist whose *Studies in the Life History of the Song Sparrow* (1937), became a classic, studied by every ornithology student for years afterward. Like many ornithologists, she was influenced by a bird book, in this case *Bird Craft* by Mabel Osgood Wright, whose color illustrations inspired young Margaret to start noticing birds and taking notes. The daughter of a professor of history at Amherst College in Massachusetts, Margaret received a B.S. degree in Biology in 1883 from Mt. Holyoke, where she also took courses in several languages. In 1915 she received a M.S. degree from Clark University, with a thesis on the food of the Northern Bobwhite or Bobwhite Quail, *Colinus virginianus*. She married a fellow graduate student, Leonard Nice, and they moved to Norman, Oklahoma, when Leonard became a professor of Physiology at the University of Oklahoma.

She took meticulous notes on the birds of Oklahoma and published *The Birds of Oklahoma*, after which she took a break from her ornithological studies to involve herself in research in the field of child psychology. She published 18 articles on language development in children, but kept her hand in ornithology, publishing notes about albino Brown-headed Cowbirds, *Molothrus ater*, winter observations of birds, the behavior of the Swainson's Hawk, *Buteo swainsoni*, as well as the nesting of Mourning Doves, *Zenaida macroura*. She also co-authored and published ornithological papers with her husband.

Margaret Morse Nice's first published papers were primarily on the abundance and occurrence of birds in various geographic locations. Later she became interested in studying bird behavior. When in 1927 her husband joined the faculty at Ohio State University in Columbus, Margaret began ornithological studies of that area while writing up the research she had done in Oklahoma. She also published many observational papers on subjects like a second mating of a robin pair, notes on Carolina Chickadees, *Poecile carolinensis*, and some observations on the birds of Europe, which led to her attending the prestigious International Ornithological Congress in Oxford. Clearly, her most significant work during the Ohio period

"The study of nature is a limitless field, the most fascinating adventure in the world."

Margaret Morse Nice

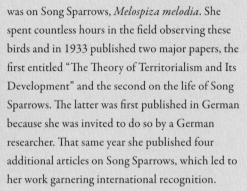

Melospiza melodia,
Song Sparrow

The 1937 "Studies in the Life History of the Song Sparrow" is one of the most well known of all bird studies.

was on Song Sparrows, *Melospiza melodia*. She spent countless hours in the field observing these birds and in 1933 published two major papers, the first entitled "The Theory of Territorialism and Its Development" and the second on the life of Song Sparrows. The latter was first published in German because she was invited to do so by a German researcher. That same year she published four additional articles on Song Sparrows, which led to her work garnering international recognition.

Margaret Morse Nice also felt a duty to inform the public about the natural world. She became a member and officer of the Columbus Audubon Society, occasionally gave nature talks, and was a frequent radio guest.

In 1937 she and her husband moved to Chicago, where she was disappointed to find that city bird life consisted mainly of House Sparrows (*Passer domesticus*). But this lack of local diversity gave her the time to do library research, review the research of others, and write up some of her previous research on Song Sparrows, the development and behavior of precocial birds, and the behavior of Brown-headed Cowbirds. For the rest of her life she continued to research, review, publish, and collaborate with others, although health problems gradually slowed her down.

Although she never held an academic or research position, Margaret earned a solid reputation in the ornithological world. Animal behaviorist and Nobel laureate Nikolaas Tinbergen wrote of her: "Through your works you have become known to ornithologists throughout the entire world as the one who laid the foundation for the population studies now so zealously pursued." Ernst Mayer, famous evolutionist, said that "she, almost single-handedly, initiated a new era in American ornithology and the only effective countermovement against the list chasing movement. She early recognized the importance of a study of bird individuals because this is the only method to get reliable life history data."

Margaret Morse Nice was honored by several professional ornithological societies, and today the Wilson Ornithological Society awards the Margaret Morse Nice medal to an outstanding researcher. She passed away in 1974 at the age of 90, a few months after her husband.

Colinus virginianus,
Northern Bobwhite (also called Bobwhite Quail)

Morse Nice's 1910 study of the Bobwhite Quail estimated that each individual eats 75,000 insects and 5 million weed seeds a year.

Tabuensis *tab-oo-EN-sis*
From Tahiti and the South Seas, as in *Porzana tabuensis*, the Spotless Crake

Tachornis *tak-OR-nis*
Greek, *tachys*, fast, and *ornis*, bird, as in *Tachornis squamata*, the Neotropical Palm Swift

Tachybaptus *tak-ih-BAP-tus*
Greek, *tachys*, fast, and *bapto*, sink, as in *Tachybaptus ruficollis*, the Little Grebe; refers to the bird's ability to compress its feathers, pushing air out so it can quickly dive underwater

Tachycineta *tak-ih-sin-ET-a*
Greek, *tachys*, fast, and *kineter*, moving, as in *Tachycineta albiventer*, the White-winged Swallow

Tachyeres *tak-ee-ER-eez*
Greek, *tachys*, fast, and *eresso*, row, as in *Tachyeres brachypterus*, the Falkland Steamer Duck, which is flightless but a fast swimmer

Tachymarptis *tak-ee-MARP-tis*
Greek, *tachys*, fast, and *marptis*, seize, as in *Tachymarptis melba*, the Alpine Swift, a fast aerial hunter of insects

Tachyphonus *tak-ee-FONE-us*
Greek, *tachys*, fast, and *phone*, sound, as in *Tachyphonus rufiventer*, the Yellow-crested Tanager; birds of this genus have rapid songs

Tachycineta albiventer,
White-winged Swallow

Tangara chilensis,
Paradise Tanager

Taczanowskii *taz-an-OW-skee-eye*
After Wladyslaw Taczanowski, Polish museum curator, as in *Cinclodes taczanowskii*, the Peruvian Seaside Cinclodes

Tadorna *ta-DORN-a*
Celtic, *tadorna*, pied waterfowl, as in *Tadorna ferruginea*, the Ruddy Shelduck

Taeniatus, -a *tee-nee-AT-us/a*
Greek, *taenia*, band or stripe, as in *Peucedramus taeniatus*, the Olive Warbler

Taeniopterus *tee-nee-OP-ter-us*
Greek, *taenia*, band or stripe, and *pteron*, wing, as in *Ploceus taeniopterus*, the Northern Masked Weaver

Taeniopygia *tee-nee-o-PIJ-ee-a*
Greek, *taenia*, band or stripe, and *puge*, rump, buttocks, as in *Taeniopygia guttata*, the Zebra Finch

Taeniotriccus *tee-nee-o-TRIK-kus*
Greek, *taenia*, band or stripe, and *trich*, hair, as in *Taeniotriccus Andrei*, the Black-chested Tyrant

Tahitica, -ensis *ta-HEE-ti-ka/ta-hee-ti-SEN-sis*
After Tahiti, as in *Hirundo tahitica*, the Pacific Swallow

Tangara *tan-GAR-a*
Brazilian Tupi native name for a brightly colored bird, as in *Tangara chilensis*, the Paradise Tanager

Tanygnathus *tan-ig-NA-thus*
Greek, *tanuo*, long, and *gnathos*, jaw, as in *Tanygnathus megalorhynchus*, the Great-billed Parrot

Tanysiptera *tan-ih-sip-TER-a*
Greek, *tanuo*, long, and *pteron*, wing, as in *Tanysiptera galatea*, the Common Paradise Kingfisher

Tarsiger *TAR-si-jer*
Greek, *tar-*, ankle, Latin, *tarsus*, a flat surface, and *ger*, bearing, carrying, as in *Tarsiger indicus*, the White-browed Bush Robin, a ground-dwelling bird

Tasmanicus *taz-MAN-ih-kus*
After Tasmania, Australia, as in *Corvus tasmanicus*, the Forest Raven

Tauraco *taw-ROCK-o*
Derived from native African name based on the bird's call, as in *Tauraco leucotis*, the White-cheeked Turaco

Tectus *TEK-tus*
Covered, as in *Vanellus tectus*, the Black-headed Lapwing

Teerinki *TER-rink-eye*
After C. G. J. Teerink, Dutch Army officer, as in *Lonchura teerinki*, the Black-breasted Manakin

Teledromas *te-le-DROM-as*
Greek, *tele*, far, and *dromas*, run, running, as in *Teledromas fuscus*, the Sandy Gallito, meaning small chicken, although it does not resemble one

Telespiza *te-le-SPY-za*
Greek, *tele*, far, and *spiza*, finch, as in *Telespiza cantans*, the Laysan Finch

Telophorus *tel-o-FOR-us*
Greek, *telo*, end, last, and *phorus*, bearer, as in *Telophorus viridis*, the Gorgeous Bushshrike

Temminckii *tem-MINK-ee-eye*
After Coenraad Temminck, Dutch ornithologist, as in *Dendrocopos temminckii*, the Sulawesi Pygmy Woodpecker

Temnurus *tem-NOO-rus*
Greek, *temno*, to cut, and *oura*, tail, as in *Temnurus temnurus*, the Ratchet-tailed Treepie

Temporalis *tem-po-RAL-is*
Tempora-, temple, as in *Ploceus temporalis*, Bocage's Weaver; refers to the temples of the head

Tenebrosa *ten-e-BRO-sa*
Tenebrae, darkness, as in *Gallinula tenebrosa*, the Dusky Moorhen

Tenuirostris *ten-oo-ee-ROSS-tris*
Tenuis, slender, and *rostrum*, bill, as in *Numenius tenuirostris*, the Slender-billed Curlew

Tephrocephalus *te-fro-se-FAL-us*
Greek, *tephros*, gray, ashy, and Latin, *cephala*, head, as in *Seicercus tephrocephalus*, the Gray-crowned Warbler

Tephrocotis *te-fro-KO-tis*
Greek, *tephros*, gray, ashy, and *otos*, ear, as in *Leucosticte tephrocotis*, the Gray-crowned Rosy-finch

Tephrodornis *te-fro-DOR-nis*
Greek, *tephros*, gray, ashy, and *ornis*, bird, as in *Tephrodornis virgatus*, the Large Woodshrike

Tephrolaema *te-fro-LEE-ma*
Greek, *tephros*, gray, ashy, and *laemus*, throat, gullet, as in *Arizelocichla tephrolaema*, the Western Greenbul

Tephronotus, -um *te-fro-NO-tus/tum*
Greek, *tephros*, gray, ashy, and *notos*, back, as in *Turdus tephronotus*, the Bare-eyed Thrush

Tanysiptera galatea, Common Paradise Kingfisher

Terenura *te-re-NOO-ra*
Greek, *tere*, soft, and *oura*, tail, as in *Terenura maculata*, the Streak-capped Antwren

Terpsiphone *terp-si-FONE-ee*
Greek, *terpsis*, enjoyment, and *phone*, a sound, as in *Terpsiphone paradisi*, the Asian Paradise Flycatcher

Terrestris *te-RESS-tris*
The ground, earth, terrestrial, as in *Zoothera terrestris*, the extinct Bonin Thrush

Tessmanni *TESS-man-nye*
After Gunther Tessman, German botanist and anthropologist, as in *Muscicapa tessmanni*, Tessman's Flycatcher

Tethys *TE-this*
Greek goddess of the sea, as in *Oceanodroma tethys*, the Wedge-rumped Storm Petrel

Tetrao *te-TRAY-o*
Greek, *tetraon*, grouse-like bird, as in *Tetrao urogallus*, the Western Capercaille, because it looks like a big grouse

Tetraogallus *te-tra-o-GAL-lus*
Greek, *tetraon*, grouse-like bird, and *gallus*, a cock, as in *Tetraogallus caspius*, the Caspian Snowcock

Tetraophasis *te-tray-o-FAY-sis*
Greek, *tetraon*, grouse-like bird, and Latin, *phasis*, pheasant, as in *Tetraophasis obscurus*, Verreaux's Monal-Partridge

Tetrax *TET-raks*
Greek, *tetraon*, game bird, as in *Tetrax tetrax*, the Little Bustard

Tetrix *TET-riks*
Greek, *tetraon*, ground-nesting bird, as in *Lyrurus tetrix*, the Black Grouse

Tetraophasis obscurus, Verreaux's Monal-Partridge

Thamnophilus palliatus, Chestnut-backed Antshrike

Teysmanni *TEZ-man-nye*
After Johannes Teijsmann (sic), Dutch botanist, as in *Rhipidura teysmanni*, the Rusty-bellied Fantail

Thalasseus *tha-LAS-see-us*
Greek, *thalassa*, sea, as in *Thalasseus maximus*, the Royal Tern

Thalassina, -us *tha-las-SEEN-a/us*
Greek, *thallasinos*, of the sea, and *hals*, sea, as in *Cissa thalassina*, the Javan Green Magpie, referring to the sea-green color of the bird

Thalassornis *tha-la-SOR-nis*
Greek, *thalassa*, sea, and *ornis*, bird, as in *Thalassornis leuconotus*, the White-backed Duck

Thamnophilus *tham-no-FIL-us*
Greek, *thamnos*, bush, and *philos*, loving, as in *Thamnophilus palliatus*, the Chestnut-backed Antshrike (see box)

Thayeri *THEY-er-eye*
After John Thayer, American ornithologist and collector, as in *Larus thayeri*, Thayer's Gull

Thinocorus *thin-o-KOR-us*
Greek, *thinos*, beach, and Latin, *corys*, lark, as in *Thinocorus rumicivorus*, the Least Seedsnipe, a shorebird found in habitats favored by larks

Thomensis *toe-MEN-sis*
From São Tomé in the Gulf of Guinea, as in *Columba thomensis*, the Sao Tome Olive Pigeon

Thoracica, -us *thor-a-SIK-a/us*
Thoracicus, breast, of the chest, as in *Poospiza thoracica*, the Bay-chested Warbling-Finch

LATIN IN ACTION

Thamnophilus palliatus, the Chestnut-backed Antshrike, inhabits thick brush, dense vines, and impenetrable thickets in South America. The male has a distinctive black crown while the female's is brown; typical of all of the species of *Thamnophilus* is the black and white pattern of the male being replaced by the brown of the female. *Palliatus* is a mantle, referring to the rufous color of the back, wings, and tail. As they forage for insects, they take slow, deliberate steps before they lunge for their prey. Disturbed by a potential predator, they will freeze motionless, sometimes for several minutes.

Thraupis THRAW-pis
Greek, a little bird, as in *Thraupis episcopus*, the Blue-gray Tanager

Threskiornis thres-kee-OR-nis
Greek, *threskos*, religious, and *ornis*, a bird, as in *Threskiornis moluccus*, the Australian White Ibis

Thripadectes thri-pa-DEK-teez
Greek, *thrips*, woodworm, and *dektes*, hunter, as in *Thripadectes ignobilis*, the Uniform Treehunter

Thripophaga thri-po-FAY-ga
Greek, *thrips*, woodworm, and *phagein*, devour, as in *Thripophaga cherriei*, the Orinoco Softtail

Thryomanes thy-ro-MAN-eez
Greek, *thruon*, reed, and *manes*, very fond of, as in *Thryomanes bewickii*, Bewick's Wren

Thryothorus thry-o-THOR-us
Greek, *thruon*, a reed, and *thorous*, rushing, leaping, as in *Thryothorus ludovicianus*, the Carolina Wren; they will inhabit marshes, but prefer woodlands and urban environments

Thula THOO-la
A region in the far north, probably snowy, as in *Egretta thula*, the Snowy Egret

Thyroideus thy-ROY-dee-us
Shield-like, as in *Sphyrapicus thyroideus*, Williamson's Sapsucker; may refer to black breast patch of female

Tibetanus ti-be-TAN-us
After Tibet, as in *Tetraogallus tibetanus*, the Tibetan Snowcock

Tibialis ti-bee-AL-is
Referring to the shin, tibia, as in *Neochelidon tibialis*, the White-thighed Swallow

Tickelli, -ae TIK-el-lye/ee-eye
After Samuel Tickell, British Army officer and ornithologist, as in *Cyornis tickelliae*, Tickell's Blue Flycatcher

Tigrina, -us ty-GRIN-a/us
Tigris, tiger, or tiger-striped, as in *Setophaga tigrina*, the Cape May Warbler

Tigriornis ty-gree-OR-nis
Tigris, tiger, or tiger-striped, and Greek, *ornis*, bird, as in *Tigriornis leucolopha*, the White-crested Tiger Heron

Egretta thula,
Snowy Egret

Tigrisoma *ty-gri-SO-ma*
Tigris, tiger, or tiger-striped, and Greek, *soma*, body, as in *Tigrisoma mexicanum*, the Bare-throated Tiger Heron

Tinamus *TIN-a-mus*
Native name from French Guinea, as in *Tinamus tao*, the Gray Tinamou

Tityra *ti-TYE-ra*
Tityrus, a character from ancient Roman poet Virgil, as in *Tityra cayana*, the Black-tailed Tityra

Tockus *TOK-us*
From Portuguese imitation of bird's call, as in *Tockus fasciatus*, the African Pied Hornbill

Todiramphus *toe-di-RAM-fus*
Todus, small bird, and Greek, *ramphos*, bill, as in *Todiramphus diops*, the Blue-and-white Kingfisher

Todirostrum *toe-di-ROSS-trum*
Todus, small bird, and *rostrum*, bill, as in *Todirostrum pictum*, the Painted Tody-Flycatcher

Todus *TOE-dus*
Small bird, as in *Todus multicolor*, the Cuban Tody

Tolmomyias *tol-mo-MY-ee-as*
Greek, *tolma*, bold, daring, and Latin, *myias*, fly, as in *Tolmomyias flaviventris*, the Ochre-lored Flatbill

Topaza *toe-PAZ-a*
Topazus, topaz, as in *Topaza pella*, the Crimson Topaz (see box)

Torgos *TOR-gos*
Greek, *torgos*, vulture, as in *Torgos tracheliotos*, the Lappet-faced Vulture

Torquata, -us, -eola *tor-KWAT-a/us/tor-kwat-ee-O-la*
Torques, twisted necklace, as in *Chauna torquata*, the Southern Screamer, with a necklace of black topped by a white one

Torquilla *tor-KWIL-la*
Torqueo, twist, turn, and *-illa*, diminutive, as in *Jynx torquilla*, the Eurasian Wryneck, named for its distinctive twisting display when threatened

Totanus *toe-TAN-us*
Italian, *totano*, moorhen, as in *Tringa totanus*, the Common Redshank

---

LATIN IN ACTION

The Crimson Topaz, *Topaza pella*, is one of over 300 species of hummingbirds. Found only in the Americas, the males have evolved a spectacular coloration with an abundance of iridescence. It is no surprise that some of their names come from the world of gems (the Amethyst-throated Mountaingem, *Lampornis amethystinus*, and the Berylline Hummingbird, *Amazilia beryllina*) or are fancifully descriptive (Rainbow Starfrontlet, *Coeligena iris*, and the Purple-throated Sunangel, *Heliangelus viola*).

Topaza pella, Crimson Topaz

Townsendi *TOWN-send-eye*
After John Townsend, American naturalist and collector, as in *Myadestes townsendi*, Townsend's Solitaire

Toxorhamphus *toks-o-RAM-fus*
Greek, *toxon*, bow, and *ramphos*, bill, as in *Toxorhamphus poliopterus*, the Slaty-headed Longbill, a small bird with a long down-curved bill

Toxostoma *toks-o-STOM-a*
Greek, *toxon*, bow, and *stoma*, mouth, as in *Toxostoma cinereum*, the Gray Thrasher, with a down-curved bill

Tragopan *TRAG-o-pan*
Greek, *tragos*, goat, and *pan*, god of the wild and flocks, as in *Tragopan caboti*, Cabot's Tragopan; the head feather tufts on the male resemble goat horns

Traillii *TRAIL-lee-eye*
After Thomas Trail, Scottish zoologist and physician, as in *Empidonax traillii*, the Willow Flycatcher

Traversi *TRA-ver-sye*
After Henry Travers, a New Zealand ornithologist, as in *Petroica traversi*, the Black Robin

Trichopsis *trik-OP-sis*
Greek, *thrix*, hair, and *opsis*, appearance, as in *Megascops trichopsis*, the Whiskered Screech Owl

Tricolor *TRIK-o-lor*
Three-colored, as in *Egretta tricolor*, the Tricolored Heron

Tridactyla *try-dak-TIL-a*
Tri-, three, and *dactylos*, toes, as in *Rissa tridactyla*, the Black-legged Kittiwake, whose hind toe is very small

Trifasciatus *try-fas-see-AT-us*
Tri-, three, and *fasciat-*, banded, as in *Carpodacus trifasciatus*, the Three-banded Rosefinch

Tringa *TRING-a*
Greek, *tringas*, a white-rumped water bird, as in *Tringa ochropus*, the Green Sandpiper

Tristigma, -ata *try-STIG-ma/try-stig-MA-ta*
Tri, three, and Greek, *stigma*, spot, as in *Caprimulgus tristigma*, the Freckled Nightjar

Tristis *TRIS-tis*
Sad, as in *Acridotheres tristis*, the Common Myna, from the Hindi word *maina*

Tristrami *TRIS-tram-eye*
After Henry Tristram, British church canon and naturalist, as in *Myzomela tristrami*, the Sooty Myzomela

Trivirgatus *try-vir-GAT-us*
Tri, three, and *virga*, stripe, as in *Accipiter trivirgatus*, the Crested Goshawk, with three bands on the tail

Troglodytes *trog-lo-DITE-eez*
Greek, *trogle*, cave, and *dytes*, dweller, as in *Troglodytes rufociliatus*, the Rufous-browed Wren; refers to its habit of disappearing into cavities and crevices when hunting for invertebrates or to rest

Trogon *TRO-gon*
Greek, *trogein*, to gnaw, as in *Trogon melanurus*, the Black-tailed Trogon; may refer to the birds' habit of gnawing into decayed trees to make a nest hole, or perhaps nibbling fruit

Tryngites *trin-JITE-eez*
Greek, *trynga*, and *-ites*, like, as in *Tryngites subruficollis*, the Buff-breasted Sandpiper, named for its similarity to sandpipers in the *Tringa* genus

Tschudii *CHOO-dee-eye*
After Johann Tschudi, Swiss explorer, as in *Ampeloides tschudii*, the Scaled Fruiteater

Turdina, -us *tur-DEEN-a/us*
Turdinus, thrush-like, as in *Schiffornis turdina*, the Brown-winged Schiffornis or Thrush-like Mourner

Turdoides *tur-DOY-deez*
Turdus, thrush, and *oides*, appearance, as in *Turdoides fulva*, the Fulvous Babbler

Turdus *TUR-dus*
Thrush, as in *Turdus merula*, the Common Blackbird

Turnix *TUR-niks*
Coturnix, quail, as in *Turnix sylvaticus*, the Common Buttonquail

Turtur *TUR-tur*
Turtle dove, as in *Turtur afer*, the Blue-spotted Wood Dove

Tympanuchus *tim-pan-OO-kus*
Tympanum, drum, and Greek, *echein*, to have, as in *Tympanuchus cupido*, the Greater Prairie Chicken; drum refers to the drumming sounds the male makes during courtship

Tyrannus, -ulus, -iscus, -ina *ti-RAN-nus/ti-ran-OO-lus/ti-ran-IS-kus/ti-ran-EE-na*
Tyrant, as in *Tyrannus albogularis*, the White-throated Kingbird

Tyto *TI-to*
Greek, *tyto*, owl, as in *Tyto capensis*, the African Grass Owl

Egretta tricolor,
Tricolored Heron

Turdus

The Common Blackbird of Europe (*Turdus merula*) and the American Robin (*T. migratorius*) are the most well-known examples of the family Turdidae, which contains about 170 thrush species in 25 genera. The true thrushes, with 65 species, are in the largest genus, *Turdus (TUR-dus)*. They are medium-sized omnivorous birds known for their melodious songs and are found on every continent except Antartica.

Pliny the Elder was a Roman author, natural historian, and philosopher who wrote *Naturalis Historia* (*Natural History*), an encyclopedia of nature. In it he gave the thrush the name *Turdus*, which survives two millennia later. The Common Blackbird's specific name, *merula*, derives from Latin and simply means blackbird; the American Robin's specific name *migratorius* means wanderer, referring to its migratory habits.

Although considered the harbinger of spring, American Robins are year-round residents throughout the US outside of Alaska and nest everywhere in North America north of Mexico. They are certainly one of the most well-known and abundant of American birds. They were named after the European Robin, *Erithacus rubecula*, which is actually a flycatcher or chat. The Common Blackbird is found throughout Europe and parts of Asia and has been introduced into Australia. It is another familiar and common bird

Turdus merula, Common Blackbird

with a population of perhaps a hundred million in Europe alone.

Thrushes often feed on the ground, eating insects, insect larvae, worms, snails, small seeds, and berries. Perhaps you have seen robins or blackbirds turning their head from side to side as they forage. They can actually hear insects crawling through the litter and worms moving in their holes.

Many true thrushes are named after locations, such as the Karoo Thrush, *T. smithi*, African Thrush, *T. pelios*, Comoros Thrush, *T. bewsheri*, Japanese Thrush, *T. cardis*, and Chinese Thrush, *T. mupinensis*. Most others are descriptive names such as the White-collared Blackbird and Bare-eyed Thrush. The former, *T. albocinctus*, Latin *albo*, white, and *cinctus*, encircling, has common and specific names that are appropriate. *T. tephronotus* (Greek *tephro*, ash-colored, and *notos*, back), the Bare-eyed Thrush, has a more apt common name than its specific name; it is gray on the back, but the bare skin around the eye is more distinctive. In the case of *T. pallidus* (Latin, *pallidus*, pale), the Pale Thrush, neither its specific nor common name is particularly descriptive.

Turdus albocinctus, White-collared Blackbird

U

Ultima UL-tee-ma
Ultimate, as in *Pterodroma ultima*, Murphy's Petrel; *ultima* seems to refer to the limited range of the bird

Ultramarina ul-tra-mar-EEN-a
Ultra, beyond, and *marina*, marine, as in *Aphelocoma ultramarina*, the Transvolcanic Jay; refers to the bird's brilliant blue coloring

Umbra UM-bra
Umbra, shade, as in *Otus umbra*, the Simeulue Scops Owl, from Simeulue Island, Indonesia

Undata, -us un-DAT-a/us
Undatus, wavelike, as in *Sylvia undata*, the Dartford Warbler

Undulata, -ua un-doo-LAT-a/un-doo-la-TOO-a
Having wavelike markings, as in *Chlamydotis undulata*, the Houbara Bustard

Unicolor oo-nee-KO-lor
Uni-, one, and *color*, color, as in *Haematopus unicolor*, the Variable Oystercatcher, which is all black, unlike many in the same genus

Unirufa, -us oo-nee-ROO-fa/fus
Uni-, one, single, *rufa*, rufous, as in *Cinnycerthia unirufa*, the Rufous Wren

Upupa oo-POO-pa
Imitation of the bird's call, as in *Upupa epops*, the Eurasian Hoopoe

Uraeginthus oo-ree-JIN-thus
Greek, *oura*, tail, and Latin, *aeginthus*, hedge sparrow, as in *Uraeginthus angolensis*, the Blue Waxbill

Uragus oo-RA-gus
Greek, *oura*, tail, and Latin, *ago*, having, as in *Uragus* (now *Carpodacus*) *sibiricus*, the Long-tailed Rosefinch

Uria oo-REE-a
Diving bird, as in *Uria aalge*, the Common Murre or Guillemot

Urichi OO-rich-eye
After Freiderich Urich, a Trinidadian naturalist, as in *Phyllomyias urichi*, Urich's Tyrannulet

Urochroa bougueri, White-tailed Hillstar

Urochroa oo-ro-KRO-a
Greek, *oura*, tail, and *khroa*, complexion, as in *Urochroa bougueri*, the White-tailed Hillstar

Uroglaux OO-ro-glawks
Greek, *oura*, tail, and *glaux*, owl, as in *Uroglaux dimorpha*, the Papuan Hawk-Owl

Uropygialis oo-ro-pi-jee-AL-is
Uropygium, rump, as in *Melanerpes uropygialis*, the Gila Woodpecker

Urosticte oo-ro-STIK-tee
Greek, *oura*, tail, and *stiktos*, spotted, as in *Urosticte benjamini*, the Purple-bibbed Whitetip

Urothraupis oo-ro-THRAW-pis
Greek, *oura*, tail, and *thraupis*, thrush, as in *Urothraupis stolzmanni*, the Black-backed Bush Tanager

Urotriorchis oo-ro-tree-OR-kis
Greek, *oura*, tail, and *triokhos*, kind of falcon or kite, as in *Urotriorchis macrourus*, the Long-tailed Hawk

Ussheri USH-er-eye
After H. B. Usher, British ornithologist, as in *Erythropitta ussheri*, the Black-crowned Pitta

Ustulatus oo-stoo-LAT-us
Burned, referring to the brownish color, as in *Catharus ustulatus*, Swainson's Thrush

Foraging

Humans are omnivores, *omni* meaning all, and *vore*, to eat, and refers to our habit of eating a wide variety of foods, plant and animal. Many birds, like crows, jays, and starlings, are omnivores, but most birds are somewhat or very restricted in their food choices because of their bill shape, digestive capabilities, or physiological needs. Obviously, long-billed shorebirds, hook-billed hawks, and flat-billed swallows and swifts are adapted to capture and eat different foods. And so we have insect/arthropod-eating birds called insectivores, fruit-eating frugivores, piscivorous fish-eaters, carnivores like hawks, and nectarivores that get their sustenance from the sugary liquid of flowers.

Birds' digestive systems have evolved to break down and incorporate food items gathered by the bill. In the winter, waxwings eat a lot of berries, some of which have a tough coating. But in sixteen minutes the berry passes through the digestive system, the coating excreted and the pulp mostly digested. The population of Myrtle Warblers (*Setophaga coronata*) in the US are so called because of their preference for Myrtle berries. These fruits are undigestible by other warblers but enable the Myrtle Warbler to winter farther north than any other North American warbler. Many tropical frugivores can eat very spicy hot peppers containing capsaicin, which is the plants' chemical defense to deter mammalian predators. Birds, however, have relatively few taste buds, enabling them to exploit food sources that other animals find distasteful. One exception is that of the Monarch butterfly whose larvae (caterpillars) eat milkweed, which contains a very distasteful alkaloid. Adult butterflies taste so incredibly bad that after one experience birds avoid Monarchs. This protects other Monarchs, as well as the non-poisonous Monarch mimic, the Viceroy butterfly. It works for a few birds as well. Some birds of the genus *Pitohui* on New Guinea eat beetles that contain a neurotoxic poison that makes the birds' skin and feathers distasteful, perhaps poisonous. This is the same poison found in the skin of poisonous frogs in Colombia.

Some birds, especially those in Corvidae, the crow and jay family, cache their food for eating later and are amazingly good at finding their secreted hoard. Nutcrackers know exactly where to dig for their nut cache even after a significant snowfall. California Scrub Jays (*Aphelocoma californica*) will bury one acorn at a time, but if they see another jay observing them they will return later to rebury the acorn elsewhere to foil any theft by the observer.

Setophaga tigrina,
Cape May Warbler

The Cape May Warbler, in order to compete successfully for food, specializes in eating insects on the tree tops.

Then there are those birds that prefer not to hunt their own food but take it from others. Kleptoparasites harass other birds and steal their food (Greek, *klepto*, I steal.) Frigatebirds often chase gulls, terns, and gannets, forcing them to drop their fish or squid in midair, where it is quickly retrieved by the frigatebird.

The survival and breeding success of birds depends in large part on the food supply and their ability to exploit it. In addition to the absolute abundance of food, competition from similar species exploiting the same food sources poses a potential problem and has resulted in an elegant solution via evolution. In any habitat, birds of different species that eat similar foods have different bill and body sizes, allowing

Ceryle rudis,
Pied Kingfisher

Most kingfishers have to divide the resources longitudinally since they are restricted to river banks or lake shores.

them to eat different sized foods. The Common and Crested Kingfishers, *Alcedo atthis* and *Megaceryle lugubris*, of central Japan share the river habitat this way, as do the very similar Sharp-shinned and Cooper's Hawks (*Accipiter striatus* and *cooperii*), the latter being one-third larger than the former.

There is the classic example of warblers in the northeastern US that forage on insects but in different parts of the trees; the Cape May Warbler, *Setophaga tigrina*, feeds on the tree tops while the while the Bay-breasted Warbler, *Setophaga castanea*, prefers the center. Insect-eating species across the world all demonstrate some version of this segregation, but the most famous example is certainly that of the Galapagos (Darwin's) finches. There are 14 finch species scattered over 19 islands, with differing sets of species on each island. The beaks of the birds vary in size depending on what other species of bird are present. Evolution has fine-tuned the size of the birds' bills in response to the neighbors' so that they make the best use of the food supply.

Accipiter cooperii,
Cooper's Hawk

The Cooper's Hawk is one of the bird species that is colonizing cities and suburbs as natural habitats disappear.

V

Validirostris *val-ih-di-ROSS-tris*
Validus, strong, and *rostrum*, bill, as in *Lanius validirostris*, the Mountain Shrike

Validus *val-EE-dus*
Strong, as in *Corvus validus*, the Long-billed Crow

Valisneria *val-is-NAIR-ee-a*
After Antonio Vallisneri, Italian naturalist, as in *Aythya valisineria*, the Canvasback

Vanellus *van-EL-lus*
Vannus, fan, and *-ellus*, little, as in *Vanellus spinosus*, the Spur-winged Lapwing

Vanga *VANG-a*
A curved blade, as in *Vanga curvirostris*, the Hook-billed Vanga

Varia, -us *VAR-ee-a/us*
Variegated, as in *Strix varia*, the Northern Barred Owl

Variegata, -us *var-ee-eh-GA-ta/tus*
Variegated, as in *Sula variegata*, the Peruvian Booby

Vauxi *VOKS-eye*
After William Vaux, American mineralogist and archeologist, as in *Chaetura vauxi*, Vaux's Swift

Velatus *vel-AH-tus*
Covered or veiled, as in *Xolmis velatus*, the White-rumped Monjita

Veniliornis *ven-il-ee-OR-nis*
In Roman mythology Venilia was a river nymph turned into a woodpecker, as in *Veniliornis maculifrons*, the Yellow-eared Woodpecker

Ventralis *ven-TRA-lis*
Ventral, belly, as in *Buteo ventralis*, the Rufous-tailed Hawk

Venusta, -us *ven-OO-sta/stus*
Venustus, beautiful, as in *Erythropitta venusta*, the Graceful Pitta

Vermiculatus *ver-mi-koo-LAT-us*
Vermis, worm, grub, wormlike (markings), as in *Burhinus vermiculatus*, the Water Thick-knee, with numerous wavy markings on the chest and back

Vermivora *ver-mi-VOR-a*
Vermis, worm, and *vorare*, to devour, as in *Vermivora cyanoptera*, the Blue-winged Warbler

Verreauxi *ver-RAWKS-eye*
After Jules Verreaux, French natural historian, and brother of Jean, as in *Coua verreauxi*, Verreaux's Coua

Verreauxii *ver-RAWKS-ee-eye*
After Jean Verreaux, French naturalist and collector, and brother of Jules, as in *Aquila verreauxii*, Verreaux's Eagle

Versicolor *ver-SIK-o-lor*
A variety of colors, as in *Amazona versicolor*, the St Lucia Amazon

Verticalis *ver-ti-KAL-is*
Crowned, as in *Cyanomitra verticalis*, the Green-headed Sunbird

Vespertinus *ves-per-TINE-us*
Of the evening, as in *Falco vespertinus*, the Red-footed Falcon

Vestiaria *ves-tee-AR-ee-a*
Vestis, cloak, *-aria*, resemble, as in *Vestiaria coccinea*, the Iiwi; its feathers were used to make robes for Hawaiian royalty

Vermivora cyanoptera, Blue-winged Warbler

Vanellus

The genus name *Vanellus* (*van-EL-lus*), from the Latin, meaning little fan, may be the name of this genus because of the way its 25 species flap their large wings in flight or act as though they have a broken wing when protecting their nest or young. These birds are called lapwings because they distract predators by dragging, flopping, or lapping one wing on the ground as if they were wounded. The approaching intruder sees easy prey in the apparently injured bird and chases after the adult, who leads the predator a safe distance from the nest and then flies away. The species name of the Senegal Lapwing, *V. lugubris*, means mournful; since mournful does not seem to describe the bird's call, the name may be from the broken-wing act.

Like many shorebirds, *Vanellus* lapwings lay around four eggs in a shallow depression. The narrow ends of the eggs face inward to prevent them from moving. Northern Lapwings commonly nest in agricultural fields but suffer 35 to 60 percent mortality from agricultural activities, including trampling by cows and predation by crows. The young, like the eggs, are cryptically colored and leave the nest very shortly after hatching and remain with the parents for 5 to 6 weeks. Because the inside of the shell is white, the parent will remove it from the nest or even bury it so as not to attract predators.

Vanellus vanellus, Northern Lapwing

The African Wattled Lapwing (*V. senegallus*) lives on plains where the only way to survey the area for predators is to find termite mounds. Perhaps the most well known of the genus is *V. vanellus*, the Northern Lapwing, because it is found across much of Eurasia. In Britain it is just called the Lapwing or, in some areas, Peewit, for its call. Up until the early twentieth century, people would collect and eat Peewit eggs. By the 1950s this practice was banned in most countries, but for many years after there was a national competition in the Netherlands to collect the first Peewit egg of the year. Finding the first egg is still a popular competition, though it is no longer taken away from the nest. Due to climate change, the first egg is being found earlier and earlier each year.

Vanellus macropterus, Javan Lapwing

The Javan Lapwing has not been reliably reported since 1940 and is most likely extinct.

Victoria, -ae *vik-TOR-ee-a/eye*
After Queen Victoria of Great Britain, as in *Goura victoria*, the Victoria Crowned Pigeon

Vidua *vy-DOO-a*
From Whydah, a town in West Africa, as in *Vidua raricola*, the Jambandu Indigobird

Vieilloti *vee-eh-LOT-eye*
After Louis Vieillot, French ornithologist and businessman, as in *Lybius vieilloti*, Vieillot's Barbet

Vigorsii *vi-GOR-see-eye*
After Nicholas Vigors, Irish secretary of the Zoological Society of London, as in *Eupodotis vigorsii*, the Karoo Korhaan

Villosus *vil-LOS-us*
Hairy, as in *Picoides villosus*, the Hairy Woodpecker

Violacea, -us *vee-o-LACE-ee-a/us*
Violaceus, violet-colored, as in *Loxigilla violacea*, the Greater Antillean Bullfinch

Virens *VIR-enz*
Becoming green, as in *Contopus virens*, the Eastern Wood Pewee

Vireo *VIR-ee-o*
Virere, to be green, as in *Vireo pallens*, the Mangrove Vireo

Virescens *vir-es-senz*
Greenish, as in *Empidonax virescens*, the Acadian Flycatcher

Virgata, -us *vir-GAT-a/us*
Virgatus, striped or streaked, as in *Sterna virgata*, the Kerguelen Tern

Sturnus vulgaris, Common Starling

Viridicata *vir-id-ih-KA-ta*
Viridius, green, as in *Myiopagis viridicata*, the Greenish Elaenia

Viridicauda *vir-id-ih-CAW-da*
Viridus, green, *cauda*, tail, as in *Amazilia viridicauda*, the Green-and-white Hummingbird

Viridicyanus *vir-ed-ih-see-AN-us*
Viridus, green, and *cyaneus*, dark blue, as in *Cyanolyca viridicyanus*, the White-collared Jay

Viridis *vir-IH-dis*
Viridus, green, as in *Merops viridis*, the Blue-throated Bee-eater

Vitellinus *vi-tel-LINE-us*
Vitellus, egg yolk (color), as in *Ploceus vitellinus*, the Vitelline Masked Weaver

Vittata, -um, -us *vit-TAT-a/um/us*
Vittatus, banded, as in *Amazona vittata*, the Puerto Rican Amazon

Vociferus *vo-SIF-er-us*
Noisy, as in *Charadrius vociferus*, the Killdeer, common name from its call

Vulgaris *vul-GAR-is*
Common, vulgar, as in *Sturnus vulgaris*, the Common Starling, reflecting its former abundance

Vultur *VUL-tur*
A vulture, as in *Vultur gryphus*, the Andean Condor

Cyanolyca viridicyanus, White-collared Jay

W

Wagleri *VAG-ler-eye*
After Johann Wagler, German herpetologist, as in *Ortalis wagleri*, the Rufous-bellied Chachalaca

Wahlbergi *VAL-berg-eye*
After Johan Wahlberg, Swedish naturalist and collector, as in *Hieraaetus wahlbergi*, Wahlberg's Eagle

Wallacii, -ei *wal-LACE-ee-eye/WAL-lis-eye*
After Alfred Russell Wallace, English naturalist, geographer, and evolutionary scientist, as in *Semioptera wallacii*, the Standardwing

Watkinsi *WAT-kinz-eye*
After Henry Watkins, British collector, as in *Grallaria watkinsi*, Watkins's Antpitta

Wetmorei *WET-mor-eye*
After Frank Wetmore, American ornithologist and paleontologist, as in *Rallus wetmorei*, the Plain-flanked Rail

Urosphena whiteheadi, Bornean Stubtail

LATIN IN ACTION

The Bornean Stubtail, *Urosphena whiteheadi*, is well described as it has a very short tail. But the tail is also wedge-shaped, hence its genus name *Urosphena* (wedge-shaped tail). Found on or near the ground in mountain forests from 2,600 to 10,000 feet (800 to 3,000 meters), it creeps through the leaf litter very surreptitiously, acting more mouse-like than birdlike, looking for insects and other invertebrates. As is typical for birds that live in dense habitats, its song and call have evolved to penetrate thick foliage with minimal attenuation.

Wetmorethraupis *wet-mor-THRAW-pis*
After Frank Wetmore, American ornithologist and paleontologist, and *thraupis*, tanager, as in *Wetmorethraupis sterrhopteron*, the Orange-throated Tanager

Whiteheadi *WHITE-head-eye*
After John Whitehead, British explorer, as in *Urosphena whiteheadi*, the Bornean Stubtail (see box)

Whitelyi, -ana *WHITE-lee-eye/ana*
After Henry Whitely, Jr, British collector, as in *Setopagis whitelyi*, the Roraiman Nightjar

Whytii *WITE-ee-eye*
After Alexander Whyte, British naturalist, as in *Crithagra whytii*, the Yellow-browed Seedeater

Whitneyi *WIT-nee-eye*
After Josiah Whitney, American geologist and explorer, as in *Pomarea whitneyi*, the Fatuhiva Monarch

Wilsonia *wil-SOWN-ee-a*
After Alexander Wilson, father of American ornithology, as in *Wilsonia* (now *Cardellina*) *pusilla*, Wilson's Warbler

Woodfordi, -ia *WOOD-ford-eye/wood-FORD-ee-a*
After Charles Woodford, Resident Commissioner Solomon Islands, as in *Nesoclopeus woodfordi*, Woodford's Rail

Xanthocephalus zan-tho-se-FAL-us
Greek, *xanthos*, yellow, and Latin, *cephala*, head, as in *Xanthocephalus xanthocephalus*, the Yellow-headed Blackbird

Xanthogaster, -tra zan-tho-GAS-ter/tra
Greek, *xanthos*, yellow, and *gaster*, belly, as in *Euphonia xanthogaster*, the Orange-bellied Euphonia

Xanthogenys zan-tho-JEN-is
Greek, *xanthos*, yellow, and *genys*, cheek, as in *Machlolophus xanthogenys*, the Himalayan Black-lored Tit

Xanthophrys zan-THO-fris
Greek, *xanthos*, yellow, and *ophrus*, eyebrow, as in *Pseudonestor xanthophrys*, the Maui Parrotbill

Xanthops ZAN-thops
Greek, *xanthos*, yellow, and *ops*, face, as in *Alipiopsitta xanthops*, the Yellow-faced Parrot

Xanthopsar zan-THOP-sar
Greek, *xanthos*, yellow, and *psar*, starling, as in *Xanthopsar flavus*, the Saffron-cowled Blackbird

Xanthopygius zan-tho-PI-jee-us
Greek, *xanthos*, yellow, and *pugios*, rumped, as in *Crithagra xanthopygius*, the Yellow-rumped Seedeater

Xanthocephalus xanthocephalus, Yellow-headed Blackbird

Xanthotis zan-THO-tis
Greek, *xanthos*, yellow, and *otis*, eared, as in *Xanthotis flaviventer*, the Tawny-breasted Honeyeater

Xantusii zan-TOOS-ee-eye
After Louis Xantus de Vesey, Hungarian collector, as in *Basilinna xantusii*, Xantus's Hummingbird

Xavieri ZAY-vee-er-eye
After Xavier Dybowski, a French explorer, as in *Phyllastrephus xavieri*, Xavier's Greenbul

Xema ZEE-ma
A word created by the namer, as in *Xema sabini*, Sabine's Gull

Xenicus ZEN-ih-kus
Greek, *xenos*, stranger, *-icus*, a suffix meaning of foreign places, as in *Xenicus gilviventris*, the New Zealand Rock Wren; at the time the bird was named, New Zealand seemed very far away

Xenopirostris zen-o-pi-ROSS-tris
Greek, *xenos*, stranger, *opsis*, appearance, as in *Xenopirostris damii*, Van Dam's Vanga

Xenops ZEN-ops
Greek, *xenos*, strange, and *ops*, face or appearance, as in *Xenops minutus*, the Plain Xenops, with a laterally flattened bill with an upturned tip

Xenus ZEN-us
Greek, *xenos*, stranger, as in *Xenus cinereus*, the Terek Sandpiper; the long upcurved bill is unusual for sandpipers

Xiphidiopicus zi-fi-dee-o-PYE-kus
Greek, *xiphidion*, small sword, and *picus*, woodpecker, as in *Xiphidiopicus percussus*, the Cuban Green Woodpecker

Xiphocolaptes zy-fo-ko-LAP-teez
Greek, *xiphos*, sword, and *colaptes*, woodpecker, as in *Xiphocolaptes albicollis*, the White-throated Woodcreeper

Xipholena zye-fo-LEN-a
Greek, *xiphos*, sword, and *olene*, arm, as in *Xipholena lamellipennis*, the White-tailed Cotinga; the white primary feathers and the slightly drooped wing posture probably account for the name

Xiphorhynchus zye-fo-RINK-us
Greek, *xiphos*, sword, and Latin, *rhynchus*, bill, as in *Xiphorhynchus pardalotus*, the Chestnut-rumped Woodcreeper

Y

Yarrellii *yar-REL-lee-eye*
After William Yarrell, English bookseller and amateur ornithologist, as in *Spinus yarrellii*, the Yellow-faced Siskin

Yaruqui *YAR-u-quee*
After Yaruqui, Ecuador, as in *Phaethornis yaruqui*, the White-whiskered Hermit

Yelkouan *YEL-koo-an*
Turkish, *yelkovan*, wind-chaser, as in *Puffinus yelkouan*, the Yelkouan Shearwater

Yemenensis *ye-MEN-ensis*
After Yemen, as in *Linaria yemenensis*, the Yemen Linnet

LATIN IN ACTION

Yucatanensis refers to the Yucatan Peninsula of Mexico, Guatemala, and Belize, a rich tropical environment with 564 known bird species, seven of which are endemic (found nowhere else in the world.) Several birds take their common names from the area, such as the Yucatan Wren, Yucatan Poorwill, Yucatan Nightjar, Yucatan Jay, Yucatan Woodpecker, and the Yucatan Flycatcher (*Myiarchus yucatanensis*). Clearly, this area is a treasure trove for insectivorous birds as there are 46 species of flycatchers in the family Tyrannidae found there, and of course many other insect eaters. The Yucatan Peninsula is also a stepping-off point for migratory flycatchers and other birds heading to North America. Many birds cross over 600 miles (1,000 kilometers) of water from the Yucatan to the US and it has to be non-stop as there are no waypoints along the route. They arrive dehydrated and exhausted and no doubt some perish, but it has worked for thousands of years, even for the tiny Ruby-throated Hummingbird, *Archilochus colubris*.

Yuhina torqueola, Indochinese Yuhina

Yersini *YER-sin-eye*
After Alexandre Yersin, Swiss bacteriologist, as in *Trochalopteron yersini*, the Collared Laughingthrush

Yncas *INK-as*
After the ancient rulers of Peru, as in *Cyanocorax yncas*, the Inca Jay, part of whose range is in the Peruvian Andes

Yucatanensis, -icus
yoo-ka-tan-EN-sis/you-ka-TAN-i-kus
After Yucatan, Mexico, as in *Myiarchus yucatanensis*, the Yucatan Flycatcher (see box)

Yuhina *yoo-HINE-a*
Nepalese, *yuhin*, as in *Yuhina torqueola*, the Indochinese Yuhina

Yunnanensis *yoo-nan-EN-sis*
After Yunnan, China, as in *Sitta yunnanensis*, the Yunnan Nuthatch

ALEXANDER WILSON
(1766–1813)

Alexander Wilson is the most well-known and highly regarded ornithologist before John James Audubon's time. Wilson, born in Paisley, Scotland, grew up poor. After leaving school at the age of thirteen to work as a weaver and cloth peddler in the Scottish countryside, he began seriously writing poetry as well. His poetry became political and he ranted against the unfair treatment of weavers by their bosses. His written tirades landed him in hot water and a short stint in prison, so he eventually saved enough money to escape to America where he expected more freedom of expression. Upon arriving at the age of 28, he picked up a gun and started shooting birds as he had done in his journeys through Scotland. After wandering around Philadelphia and working at weaving, peddling, and printing, he finally found a post as a schoolteacher.

Wilson became acquainted with William Bartram, a naturalist and talented artist, who drew botanical and ornithological subjects. Wilson borrowed some of Bartram's paintings and learned to illustrate by copying them. He left his teaching job and took on the task of revising a 22-volume encyclopedia. This job provided a nice salary and connections in the publishing world. He decided on a project to describe every bird in America, an ambitious effort for someone whose artistic talents were still evolving and whose knowledge of American birds was just developing. As Wilson wrote to Bartram: "I dare say you will smile at my presumption when I tell you that I have seriously begun to make a collection of drawings of the birds to be found in Pennsylvania, or that occasionally pass through it: twenty-eight, as a beginning, I send for your opinion."

Scottish-born Alexander Wilson became a highly regarded ornithologist and author/illustrator after emigrating to the US in 1794.

Even though he shot many birds and kept many live specimens, he still needed help in identification, but his passion and work ethic convinced his publisher to accept his proposal for a multiple-volume series called *American Ornithology*. There was one major condition: that Wilson sell subscriptions to pay for it.

After Volume I was produced, Wilson walked and rode horseback thousands of miles trying to sell subscriptions. He slept in the woods and in Indian villages, fending off a variety of hardships and illnesses. As he traveled, he observed and shot birds, collecting perhaps two-thirds of the species east of the Mississippi River. In 1810 at the age of 44, in Louisville, Kentucky, he tried to sell the plates to a storekeeper who apparently admired the work and was ready to subscribe but ultimately turned down Wilson after conferring with the store's senior business partner. The storekeeper, 25-year-old John James Audubon, preferred to be outside shooting birds and drawing them himself rather than reading about them. A story has it that the senior business partner whispered to Audubon in French

(Audubon was born in Haiti) that his (Audubon's) art was much better than Wilson's.

Audubon later claimed that he lent Wilson several paintings. This may have been a ploy to hide what many claim is plagiarism on Audubon's part. Several of Audubon's birds are very close copies of Wilson's works, and a few dozen show a distinct likeness.

From 1810 to 1812 Wilson completed a total of six volumes of his book. He made copperplate etchings that were composed of simple black lines on white paper. All the coloration had to be done by hand with watercolors. So each page, even though a "reproduction," was an original work of art. He tried to find artists to help him with this work but most did not meet his standards so he ended up doing most of the painting himself.

As Wilson traveled, he cultivated some aristocratic and moneyed friends who supported his endeavors and facilitated his travels. But overwork, constant travel, and various illnesses finally took their toll; he died in 1813 at the age of 47, some say in pursuit of a bird across a river. His legacy was the nine-volume *American Ornithology* (1808–1814), which illustrated 268 species of birds, 26 of which had not previously been known. This magnum opus made him known as the Father of American Ornithology.

The Wilson Ornithological Society was founded in 1888 and publishes a quarterly journal, *The Wilson Journal of Ornithology*. The WOS recognizes the important role of the serious amateur in ornithology.

Several species of birds were named after Alexander Wilson, including Wilson's Storm-petrel, *Oceanites oceanicu*; Wilson's Plover, *Charadrius wilsonia*; Wilson's Phalarope, *Phalaropus tricolor*; Wilson's Snipe, *Gallinago delicata*; and Wilson's Warbler, *Cardellina pusilla*.

Phalaropus tricolor, Wilson's Phalarope

Phalarope means coot-footed, describing the lobes on the bird's foot that help it when walking on muddy ground and swimming.

"Particular species of birds, like different nations of men, have their congenial climes and favorite countries; but wanderers are common to both; some in search of better fare, some of adventures, others led by curiosity, and many driven by storms or accident."

Alexander Wilson, American Ornithology, or The Natural History of the Birds of the United States

Z

Zambesiae *zam-BEEZ-ee-ee*
After the Zambesi River, Africa, as in *Prodotiscus zambesiae*, the Green-backed Honeybird

Zantholeuca *zan-tho-LOY-ka*
Greek, *xantho*, yellow, *leukos*, white, as in *Erpornis zantholeuca*, the White-bellied Erpornis

Zaratornis *zar-a-TOR-nis*
After Zarate, Argentina, and *ornis*, bird, as in *Zaratornis stresemanni*, the White-cheeked Cotinga

Zavattariornis *za-vat-tar-ee-OR-nis*
After Edoardo Zavattari, Italian zoologist and explorer, and *ornis*, bird, as in *Zavattariornis stresemanni*, Stresemann's Bushcrow

Zebrilus *ze-BRIL-us*
French, *zebre*, zebra, and *-ilus*, diminutive, as in *Zebrilus undulatus*, the Zigzag Heron

> **LATIN IN ACTION**
>
> The White-throated Sparrow is aptly named as *Zonotrichia albicollis*, the white-striped or zoned small bird with a white neck. There are two populations, one with a white crown and one with a tan crown. Males of both crown colors prefer white-crowned females, but females of both crown colors prefer males with tan striped crowns. So both variations will continue to exist.

Zeledonia *ze-le-DON-ee-a*
After Jose Zeledon, Costa Rican naturalist and collector, as in *Zeledonia coronata*, the Wrenthrush

Zenaida *zen-EH-da*
After Princess Zenaide Bonaparte, as in *Zenaida auriculata*, the Eared Dove

Zimmeri, -ius *ZIM-mer-eye/zim-MARE-ee-us*
After John Zimmer, American ornithologist, as in *Scytalopus zimmeri*, Zimmer's Tapaculo

Zonerodius *zo-ne-RO-dee-us*
Greek, *zone*, band, and *erodios*, heron, as in *Zonerodius heliosylus*, the Forest Bittern

Zonotrichia *zo-no-TRIK-ee-a*
Greek, *zone*, band, and *trichias*, small bird, as in *Zonotrichia albicollis*, the White-throated Sparrow (see box)

Zoothera *zoo-o-THER-a*
Greek, *zoon*, animal, and *theros*, hunter, as in *Zoothera dixoni*, the Long-tailed Thrush

Zosterops *ZOS-ter-ops*
Greek, *zoster*, girdle, and *ops*, appearance, as in *Zosterops senegalensis*, the African Yellow White-eye

Zonotrichia albicollis,
White-throated Sparrow

Zosterops

Zosterops (*ZOS-ter-ops*) means girdle eye, from the Greek *zoster*, girdle, and *ops*, eye. Their common name of white-eye or speirops (Greek *speira*, circle, and *ops*) aptly describes the birds of this genus, with their wide ring of white feathers around the eyes. There are 98 species of *Zosterops*, one of the largest genera in the bird world. They live in regions of sub-Saharan Africa, Asia/Indonesia, and Australasia. Once thought to be related to nectar-feeding birds like honeyeaters because of their brush-like tongue, recent DNA evidence puts them closer to Old World warblers.

These small birds, only 3–4 inches (10–12 centimeters) long and 0.3–0.5 ounces (10–12 grams) in weight, have been very successful in a variety of habitats, climate zones, and altitudes. They are very good colonizers and easily invade disturbed habitats. The greatest number of white-eyes is found in the Solomon Islands, which are home to 11 species, but only one or two species exist on each island.

White-eyes are very social, congregating in flocks as they move through habitats in search of food while constantly calling to each other. Members of the flock apparently develop close relationships; ringing/banding studies have captured many of the same birds in a flock year after year. Flocks are sometimes small, but up to 500 birds have been counted in a single group.

One of the reasons for white-eyes' success is their ability to enter a state of torpor at night, dropping their body temperature about 41ºF (5ºC), resulting in a halving of their metabolic rate. At dusk white-eyes gather in small groups, but as darkness approaches, these small groups coalesce into a larger group. In their roost, they perch so close together that the wings and tails of neighboring birds often overlap. Their need for social interaction is so strong that they accept birds of other species, even other families, in their group. One evening roosting group in Thailand numbered over 1,000 birds!

The Japanese White-eye, *Z. japonicas*, native to Asia and the Far East, has been introduced elsewhere as a pet and for insect control, but has become a pest itself. It is now the most common land bird in Hawaii.

Several species of white-eyes are threatened by habitat destruction and the invasive Red-whiskered Bulbul, *Pycnonotus jocosus*, that preys on their eggs.

Zosterops kikuyuensis, Kikuyu White-eye

Zosterops ficedulinus, Principe or São Tomé White-eye (pictured left)

Islands in the Gulf of Guinea host *Zosterops ficedulina* on Principe and *Zosterops feae* on São Tomé but they may be the same species.

Glossary

Binomial
The scientific name that consists of two words, the genus and species

Breast
Area of the bird between the neck and the abdomen

Carina
Also known as the keel, the ventral extension of the sternum (breastbone) that serves as the anchor for the breast muscles used in flight

Cere
Latin for wax, a waxy structure that covers the base of the upper bill and usually contains the nares

Covert
A type of feather that covers the flight and tail feathers (or the base of them) and the ears

Crest
An extension of crown feathers above the head, either fixed or moveable

Crown (or cap)
The top of the head

Culmen
The upper ridge of a bird's beak

Decurved
Downcurved, curving downward

Endemic
Native or restricted to a certain country or area

Extant
Still living; not extinct

Family
The taxonomic category above genus; contains one or more genera

Foraging
The behavior involved in finding food

Furcula
Dim. of *furca*, fork, fused clavicle bones that help anchor the breast muscles; the wishbone

Gastroliths
Small stones ingested and stored in the stomach to help grind food

Genera
Plural of genus

Genus
The category above species in the taxonomic hierachy; contains one or more species

Holotype
The single specimen designated as the type for naming a species

Lamellae
Strainer-like projections of the bill edges found in some waterfowl

Lobe
A rounded projection of a body part as lobes on the toes of a foot

Malar
Cheek area

Mandible
The upper and lower part of the bill; half (usually lower) of a jaw

Nape
Back of the neck

Nares
Nostrils

Onomatopoeia
Words like chachalaca, hoopoe, or cuckoo that imitate or suggest the sounds associated with the bird

Oology
The study of eggs

Operculum
A flap of tissue that covers the nares in some birds

Orbit
Cavity in the skull that contains the eye

Order
The category above family in the taxonomic hierarchy that contains one or more families

Ornithologist
A scientist who studies birds (*orni*, bird, and *ology*, the science of)

Palmate
Having a shape similar to a hand; digits all extending from the same point

Pelagic
Ocean going, feeding on the ocean

Plumage
The layer of feathers that covers the bird and the arrangement, color, and pattern of those feathers

Primary feathers
Wing feathers attached to the hand of the bird and used for propulsion

GLOSSARY

Pygostyle
Fused caudal vertebrae to which the retrices (tail feathers) are attached; colloquially, the pope's (or parson's) nose

Ramphotheca
The keratin covering of the jaws; the outer covering of the beak

Recurved
Curved upward

Remige
Feathers of the arm involved in flight (i.e. propulsion and lift)

Retrices
Feathers of the tail

Rictal bristle
Modified feathers at the corner of the mouth with a tactile sense

Rostrum
The beak or bill

Rump
The area of the bird where the tail meets the back of the bird

Scientific name
The binomial or trinomial name consisting of genus, species, and sometimes subspecies

Secondary feathers
Wing feathers attached to the ulna and used for lift

Semipalmate
Partly palmate; toes partly webbed

Species
The basic unit of taxonomic classification; a group of organisms capable of interbreeding and producing viable offspring

Specific epithet
Refers to the species part of the scientific name

Superciliary
Over the eye

Syndactyl
Two or more digits fused together (*syn*, together, and *dactyl*, finger)

Systematics
The study of the relationships of living things

Taxonomy
The science of classification and naming

Tertiary Feathers
Short, innermost flight feathers of the wing that primarily serve to cover the gap between the wing and the body in flight

Trinomial
The scientific name that consists of three words: the genus, species, and subspecies

Uropygial gland
The gland at the base of the tail that produces oil; also called the preen gland

Vent
The common opening for waste products

Zygodactyl
Two digits forward and two back (*zygo*, yoke, and *dactyl*, finger)

Campylopterus largipennis, Gray-breasted Sabrewing (p. 21)

Bibliography

Adler, Bill (ed.). *The Quotable Birder*. New York, New York. The Lyons Press, 2001.

Arnott, W. Geoffrey. *Birds in the Ancient World from A to Z*. Oxford, England. Routledge, 2012.

Ayers, Donald M. *Bioscientific Terminology*. Tucson, Arizona. The University of Arizona Press, 1972.

Beolens, Bo and Watkins, Michael. *Whose Bird?* New Haven, Connecticut and London, UK. Yale University Press, 2003.

Bird, David M. *The Bird Almanac*. Buffalo, New York, Firefly Books, 1999.

Clements, James F. *The Clements Checklist of Birds of the World* (Sixth Edition). Ithaca, New York. Cornell University Press, 2007.

Dorsett, R. J. Philip Alexander Clancy, 1917, *Ibis* 144 (2), 369-370, 2002

del Hoyo, Josep, Elliott, Andrew, Sargatal, Jordi (eds. vol. 1–7), and Christie, David A. (ed. vol. 8–16). *Handbook of Birds of the World*. Barcelona, Spain. Lynx Edicions, 1992–2011.

Ehrlich, Paul R., Dobkin, David S. Dobkin, and Wheye, Darryl. *The Birder's Handbook*. New York, New York. Simon and Schuster, 1988.

Gill, Frank B. *Ornithology* (Third Edition). New York, New York. W. H. Freeman and Co., 2007.

Gotch, A.F. *Latin Names Explained*. London, UK. Cassel and Company, 1995. Gill, F & D Donsker (Eds). 2013. IOC World Bird List (v 3.5). doi : 10.14344/IOC.ML.3.5

Gould, John. *The Birds of Great Britain*, London, UK. Taylor and Francis, 1873.

Gruson, Edward S. *Words for Birds*. New York, New York. Quadrangle Books, 1972

Harrison, Lorraine. *Latin for Gardeners*. Chicago, Illinois. University of Chicago Press, 2012.

Hill, Jen (ed.). *An Exhilaration of Wings*. New York, New York. Viking Penguin/Penguin Putnam, 1999.

Jobling, James A. *Helms Dictionary of Scientific Bird Names*. London, UK. Christopher Helm (A&C Black), 2010.

Moorwood, James. *A Latin Grammar*. Oxford, UK. OUP, 1999.

Rosenthal, Elizabeth J. *Birdwatcher: The Life of Roger Tory Peterson*. Guilford, Connecticut.The Globe Pequot Press, 2008.

Sibley, David Allen. *The Sibley Guide to Birds*. New York, New York. Alfred A. Knopf, 2000.

Sibley, C. G. and Monroe, B. L. *Distribution and Taxonomy of Birds of the World*. New Haven, Connecticut. Yale University Press, 1990.

Watts, Niki. *The Oxford New Greek Dictionary*. New York, New York. The Berkeley Publishing Group, 2008.

Weidensaul, Scott. *Of a Feather*. Orlando, Florida. Houghton-Mifflin Harcourt, 2007.

Websites

English-Word Information
http://www.wordinfo.info

IOC World Bird List
http://www.worldbirdnames.org

IUCN 2012. The IUCN Red List of Threatened Species. Version 2012.2.
http://www.iucnredlist.org

LatDict, Latin Dictionary on the web
http://latin-dictionary.net/search/latin/caudata

MyEtymology
http://www.myetymology.com/

Online Etymology Dictionary
http://www.etymonline.com

Credits and Acknowledgements

Picture Credits

p. 27, bottom © Dorling Kindersley | Getty Images
p. 46 © De Agostini | Getty Images
p. 47, top © Getty Images
bottom © Linda Hall Library
p. 52, top © Encyclopaedia Britannica | UIG | Getty Images
p. 54, top © Encyclopaedia Britannica | UIG | Getty Images
p. 58 © Dorling Kindersley | Getty Images
p. 59, top © Dorling Kindersley | Getty Images
p. 67, top © Hein Nouwens | Shutterstock.com
p. 76 © Tony Wills | Creative Commons
p. 77, top © De Agostini | Getty Images
bottom © Hein Nouwens | Shutterstock.com
p. 90 © De Agostini | Getty Images
p. 94 Red-shouldered Vanga © H. Douglas Pratt
p. 105 © De Agostini | Getty Images
p. 120, bottom © De Agostini | Getty Images
p. 146 © DEA PICTURE LIBRARY | Getty Images
p. 150 © Tony Wills | Creative Commons
p. 154 © Max Planck Gesellschaft | Creative Commons
p. 155, top © DEA PICTURE LIBRARY | Getty Images
bottom © karakotsya | Shutterstock.com
p. 170, botom © De Agostini | Getty Images
p. 190 © De Agostini | Getty Images
p. 191, top © Dorling Kindersley | Getty Images
bottom © Hein Nouwens | Shutterstock.com
p. 198 © Time & Life Pictures | Getty Images
p. 199, top © Encyclopaedia Britannica | UIG | Getty Images
p. 206, top © DEA PICTURE LIBRARY | Getty Images
p. 211, top © Duncan Walker | Getty Images

All images in this book are public domain unless otherwise stated.

Every effort has been made to credit the copyright holders of the images used in this book. We apologise for any unintentional omissions or errors and will insert the appropriate acknowledgment to any companies or individuals in subsequent editions of the work.

Acknowledgments

We would like to thank James Evans at Quid Publishing for working with us on the initial phases of the book. Via numerous emails, we were led, encouraged, stimulated, and in some cases, politely demanded of, by Lucy York who kept our proverbial feet to the fire and noses to the grindstone. She also did a yeoman's job of editing and proofreading a very detailed tome. Ian Carter gave the work another fine-toothed comb-through with his detailed eye for editing as well as ornithological expertise.

We also offer our thanks to Bo Boelens of FatBirder.com, probably the best ornithological website on the internet, for his freely given invaluable advice on birds and books.

And we would like to thank our high school Latin teachers, who had a major influence on us in our careers and pleasures. We did not realize the benefit at the time, but such is the risk and value in educating the young.

Roger Lederer and Carol Burr

Bombycilla garrulus,
Bohemian Waxwing (p. 33)

Passerina caerulea,
Blue Grosbeak (p. 39)